Wireless:
Strategically Liberalizing
the Telecommunications Market

TELECOMMUNICATIONS
A Series of Volumes Edited by Christopher H. Sterling

Bracken/Sterling • Telecommunications Research Resources: An Annotated Guide

Brock • Toward a Competitive Telecommunication Industry: Selected Papers from the 1994 Telecommunications Policy Research Conference

Brock/Rosston • The Internet and Telecommunications Policy: Selected Papers from the 1995 Telecommunications Policy and Research Conference

Lehr • Quality and Reliability of Telecommunications Infrastructure

Mody/Bauer/Straubhaar • Telecommunications Politics: Ownership and Control of the Information Highway in Developing Countries

Noll • Highway of Dreams: A Critical View Along the Information Superhighway

Regli • Wireless: Strategically Liberalizing the Telecommunications Market

Teske • American Regulatory Federalism and Telecommunications Infrastructure

Williams/Pavlik • The People's Right to Know: Media, Democracy, and the Information Highway

Wireless:
Strategically Liberalizing the Telecommunications Market

Brian J. W. Regli

Routledge
Taylor & Francis Group

NEW YORK AND LONDON

First published 1997 by Lawrence Erlbaum Associates,Inc., publishers

Published 2010 by Routledge
605 Third Avenue, New York, NY 10017
4 Park Square, Milton Park, Abingdon, Oxon OX14 4RN

*Routledge is an imprint of the Taylor & Francis Group,
an informa business*

Cover design by Jennifer Sterling

Library of Congress Cataloging-in-Publication-Data

Regli, Brian
Wireless : strategically liberalizing the telecommunications
market / Brian Regli.
p. cm.
Includes bibliographic references and index.
ISBN 0-8058-2581-9 (c : alk. paper). — ISBN 0-8058-2582-7
(p: alk. paper)
1. Cellular radio—Case Studies. 2. Wireless
communication systems—Case studies. 3.
Telecommunication policy—Case studies. 4.
Competition—Case studies. I. Title.
HE9713.R44 1997
384.5—dc21   97-5611
                                                    CIP

ISBN 13: 978-0-8058-2582-4 (pbk)

The final camera copy for this work was prepared by the author
and, therefore, the publisher takes no responsibility for the
consistency or correctness of typographical style. However, this
arrangement helps to make publication of this kind of scholarship
possible.

Contents

Preface

This is a book about how to use wireless access as a tool to further competition and sustainable development in the telecommunications sector. The focus of the book is mainly on the potential impact of two-way wireless communications, through existing service models such as cellular, personal communications services/networks (PCS/PCN), and emerging applications such as wireless local loop. At the same time, the discussion also includes other important forms of wireless access, such as satellite, enhanced specialized mobile radio (ESMR), and paging systems.

The research presented is a combination of academic studies, policy analysis, and corporate strategic thinking about the future of the telecommunications industry. The goal has been to create a book that would be of value to managers in both the private and public sector interested in understanding the present and future course of global privatization, liberalization policy, and the potential role of wireless access.

For policymakers and academic researchers, the value of this book comes from the discussions of how wireless access is evolving and what its potential impact will be on regulatory regimes and government institutions. By presenting both the theory and the comparative business context for the growth of wireless access around the world, the book identifies a common opportunity and potential response for policymakers in both the developed and developing world: furthering the liberalization of the telecommunications market through a specific focus on expanding competition in wireless access. Specific comparisons of competitive conditions and regulations in the United States, United Kingdom, Russia, and Brazil offer a variety of conditions from which general principles on sector performance and growth can be drawn.

For corporate managers, the book presents a full range of options that companies may wish to consider as they position themselves as competitive wireless service providers in both the developed and developing world. Marketing, managing, and positioning corporate investment among business and political leaders is a key focus of discussion, and the four individual country case studies include suggestions for corporate strategy appropriate to each political and business environment.

To provide as complete a picture as possible of the approaches to infrastructure development and their implications in the real world of business and service provision, I have drawn from a wide range of disciplines. As is often the case with a cross-functional approach to research, there are points where the depth of discussion may seem insufficient for a specialist in a particular portion of the field; in those areas, I have attempted to demarcate the limits of the analysis and suggest further direction and grounding.

ACKNOWLEDGMENTS

Inevitably, I have relied on the patient advice and mentoring of professionals from various academic and corporate disciplines who have helped me draw this material together. In particular, I would like to thank Scott Thompson of the Fletcher School of Law and Diplomacy for his help on development theory, Stuart Brotman for his contributions on telecommunications law and regulation, a whole range of people for their background on wireless communications, including Terrence McGarty, Winn Schwartau, Bob Ratcliffe, and the members of Coopers & Lybrand's Information and Telecommunications Industry Group, in particular William Cobourn (now of Price Waterhouse), James Murphy and Richard Gnospelius. On the development of the Russian telecommunications sector, I have to thank Mikhail Kazachkov of the Freedom Channel and Shawn O'Donnell of the Massachusetts Institute of Technology for the opportunity to work with them on a variety of projects in Russia over the past few years, and to Lee McKnight of MIT's Center for Science, Technology, and Policy and Antonio Botelho for their ongoing education on the Brazilian telecommunications market.

I also owe a debt of gratitude to Diana Shayon, Reena Gordon and Kathleen Brust of HRN, the management and communications consultancy where I have worked for the past 3 years, for the opportunity to serve clients like NYNEX, Bell Atlantic, BellSouth, Ameritech, and Telstra. My thanks also to Chris Sterling, editor of the LEA Telecommunications Series, whose comments and criticisms have greatly strengthened this work, and without whose support this project would have never gotten past the initial stages.

Finally, to Russ Neuman of the Kennedy School of Government, for the patience, generosity, and tireless effort that only a true mentor can provide. And to my family for their support, especially my father-in-law Paul Witte for his initial editing of the manuscript, and my wife Tanya, who has put up with too many sleepless nights listening to me clicking away at my computer.

—*Brian J. W. Regli*

Competition and Development in the Telecommunications Sector

It's a jungle out there.

It used to be a pretty well kept garden, perhaps overgrown a bit with weeds in some corners and underfertilized in others, but at least there was a sense that things were under control in the telecommunications sector. Governments owned the networks, provided the services, and kept the system running most of the time in the developed world, some of the time in poorer countries. Competition did not exist, infrastructure development was tied to public sector budget priorities or regulations, and the telecommunications sector provided the best investments for widows and orphans that Wall Street had to offer.

But today's global telecommunications sector looks more and more like a jungle each day.[1] The organizational boundaries of the telecommunications sector shift on a daily basis. From internal reorganizations to external alliances and mergers, a pattern of evolution has been set in motion by the policies of privatization and liberalization worldwide. The species and creatures of this jungle are mutating and now provide new kinds of "features" and "service offerings." The "biodiversity" of the telecommunications jungle is rapidly evolving as well, moving quickly away from the undifferentiated substance of plain old telephone service. In fact, there are almost as many theories about how to use the rich natural resources of this competitive jungle as there are new technologies, companies, and government regulators emerging on the scene.

That dynamic of change has irrevocably altered the path of telecommunications development and requires a reexamination of the sustainability of telecommunications infrastructures in this new environment. In this regard, the analogy to the sustainability of the jungle is apt and timely. The rainforest, much like a telecommunications network, is a resource. People draw on that resource for the purposes of development. Given the right conditions, the ecosystem can thrive and provide riches for those who protect the forest's sustainability. But what are the right conditions for this rainforest known as the telecommunications sector?

1

To take our cues from the language of sustainable development pioneered by the environmental economists and policymakers, what we are looking for is a process for leveraging existing resources while simultaneously protecting them for future use. The art is balance. We need to give back to the forest what we take out, or eventually the forest will no longer exist. In many ways, sustainable development for the telecommunications sector means the same thing as for a tree in a tropical rainforest: Given a sufficient amount of sunlight and water (investment), the appropriate soil (regulation), and a community to feed from and compete with (the market), the trees and various inhabitants of the jungle (telecommunications companies) should grow and thrive.

But then humans come along and mess it all up. Slash and burn policies take away from the viability of institutions. New technologies are put on the shelf by companies who profit from existing technologies. Regulators, afraid of losing their jobs to the give and take of market competition, fence off territory and restrict activity. Thinking globally and acting locally takes on a whole new meaning as public and private sector institutions pull resources from the vast ecosystem that is beginning to thrive in our midst.

This book was written to try to make sense of the various strategies that connect the development of the telecommunications sector to successful corporate strategy and good public policy. Many commentators from academic, corporate and public sector circles have drawn connections between advanced telecommunications technology and the potential for economic growth, political participation, and social development. Increased competition has shifted some of the arguments made to guide government regulation and corporate investment policy, but the themes have largely remained the same. All around the world, corporate and public sector managers are still searching for the right combination of competition and development appropriate to the needs of today's customers and tomorrow's global information society.

Interestingly enough, one of the newer creatures in this jungle may be able to fight off the predators that roam largely unchecked through the underbrush—and provide a foundation for sustainability. Wireless communications, such as cellular, satellite, and wireless local loop technologies, have been developed to provide personal access to the telecommunications infrastructure. Of all the creatures of the modern telecommunications jungle, wireless communications has shown the energy to be a change agent for telecommunications providers and the potential to serve as a focal point for investment in the "public good" of service provision.

Yet wireless access has remained largely an appendage to the existing wireline telecommunications network. Dominant providers have been compelled by regulators to separate out wireless services from wireline, ensuring that providers cannot cross-subsidize, but denying them the opportunity to provide the seamless communications access customers want. High-cost applications are being used to sustain revenue streams, while lower-cost applications remain largely on the drawing board or relegated to smaller markets or test beds. That is no way to promote the evolution to competition and diversity in global telecommunications.

This book suggests one potential alternative to the present evolutionary course for wireless access, a policy called strategic liberalization. Strategic liberalization is defined as the implementation of specific policy measures to increase competition in the market for wireless access services, such as cellular, wireless local loop, and satellite communications. It argues that a strategic focus on wireless communications will provide a sustainable foundation for the continued growth of the telecommunications sector and the successful introduction of new products and services in an environment of facilities-based competition.

But why should public and private sector managers place a strategic focus on wireless communications, as opposed to the variety of other possible access technologies and service offerings that may be made available in an increasingly competitive market? Why should regulators think of wireless communications as more than just an ancillary service to the traditional wireline network, and understand its potential as a cornerstone for telecommunications development? Those are just the opening questions in this comparative analysis of infrastructure development strategies and discussion of the potential role for new wireless telecommunications technologies throughout the world.

TELECOMMUNICATIONS DEVELOPMENT: CORPORATE AND GOVERNMENT INSTITUTIONS

May you live in interesting times, the old Chinese curse goes. For corporate managers and government policymakers throughout the world, our interesting times might certainly be seen more as a curse than as a boon. The institutions that have been constructed to provide telecommunications and information services to the people of the world are under increasing pressure. The pressure from customers is evident in both the developing and developed world: Improve the quality of existing services, lower the cost, and make advanced services more accessible to everyone. At the same time, new technologies are quickly making parts of the traditional telecommunications network obsolete. The development and introduction of new services requires a further institutional transformation for the world's private and public telecommunications providers. So there are forces driving the transformation of public and private institutions in the sector, affecting both the nature of demand for telecommunications and information services and the means by which these services are supplied.

It has not always been this way. Since the inception of public telephone and telegraph service in the middle of the 19th century, the primary models for the development of national and global telecommunications networks have been dictated by economies of scale and national interests. Centralization and monopoly have been the key, resolving the problems of both supply and demand in a neat, tightly regulated package. But we live in a world that can no longer sustain that paradigm of development, and the well rehearsed debate about the importance of economic intervention by government and the value of free-market policies offers limited direction in these interesting times.

That is because the institutional foundation of corporate service providers and government regulators has shifted dramatically in the past few decades. In turn, the grounds for collective action between public and private sector institutions has also shifted. There is a need to take a step back and redefine how we understand the institutions responsible for telecommunications development and the grounds on which a common framework for telecommunications development can be set.

The Transformation of Corporate Institutions in the Telecommunications Sector

To put it simply, corporate institutions used to be public utilities, either owned by governments directly in single shareholder arrangements or highly regulated and closely watched. The justifications for this approach were based on assessments of the economies of scope and managerial needs of telecommunications providers; high fixed costs made new entry difficult, and the duplication of infrastructure was considered a wasteful investment. Institutional structures were based on those assessments, and highly bureaucratized, highly centralized models for telecommunications development were born.

But the technological foundation for providing telecommunications services has been altered by the development of digital switching, transmission technology, computerization, and a host of related innovations. New providers found that the barriers for entry were not as high as once thought. Customers demanded new kinds of services, new kinds of connections and new capabilities that the old copper telephone network could not really support.

The result has been the uneven adaptation of telecommunications providers throughout the world. On one hand, they remain closely tied to their heritage as the main driver of telecommunications investment and as the institution responsible for providing service to as many people as possible at a price palatable to the political institutions. On the other hand, providers are attempting to carve out a vision of themselves as competitive and agile, able to respond to customer demands, new technologies, and marketplace changes. Anyone working in the telecommunications industry around the world can tell you that they are not quite there yet, but the rhetoric will have to meet the reality as competition presses for more institutional changes.

The Transformation of Government Institutions

Government institutions once had the comfortable position of oversight alone, playing the roles of customer advocate, business economist, financial analyst, social theorist, and infrastructure planner all at the same time. In some cases, such as in the European countries, the government institutions responsible for managing and regulating the telecommunications infrastructure were highly centralized.

In the United States, competing local and state jurisdictions mingled with what was a wholly American solution to the need for a strong private sector and a highly regulated market: the creation of AT&T. In developing countries, governments often spent a great deal of time trying to figure out how to drain money from telecommunications revenues to meet social needs.

Now the demands on government institutions are more varied. Government institutions are now being called on to act as investment bankers to provide financing, as arbiters in disputes between competitive adversaries, as repositories of knowledge and capabilities, and as identifiers of new technological developments. But there are few, if any, models for this kind of proactive governmental institution, and there are many voices calling for the outright abolition of regulatory agencies and government institutions involved in the telecommunications sector. In theory, a highly competitive market should require less regulation, not more. But because there are jobs at stake among professional regulators and government bureaucrats, there will be an increasingly apparent effort to define roles where government institutions can play a credible and active role in telecommunications development.

The Grounds for Common Action

Common action among government and corporate institutions defines the ability of nations to actively determine their progress toward social goals and objectives, such as economic growth, and modernization. A combination of legal structures, bureaucratic arrangements, and social compacts guide institutional interaction which, in turn, affects how resources are allocated and goals are set. The effort to use corporate and public policy to jointly fuel economic and political development depends on institutional structures and arrangements.

The framework shared over the past century was one that had relatively straightforward accountabilities. The political goals were increased penetration and access for citizens; the corporate goals were modified accordingly, often to the detriment of profitability. But that arrangement is no longer sustainable, because governments are aware that the mandates of the past are not relevant to the market of the future, and corporations know that sustainability in a competitive environment is far different from treading water in a highly protected market.

The problem is that, in an environment where change is constant and no one is quite sure what the purpose of these institutions is going to be, establishing a ground for common action is extraordinarily difficult. Changes in technological, economic, and social facts open new opportunities to improve how a community allocates resources and sets goals, and our global society should be looking to identify the opportunities and take advantage of them.

That is not to say there has been no collective response to these opportunities. In fact, the response has been dramatic, not incremental, even though the results have been uneven and more questions have been raised than have been answered. The response can be summed up in two words: privatization and liberalization.

PRIVATIZATION AND LIBERALIZATION:
THE RESPONSE TO CHANGE

The common agenda that has been defined and enacted over the past 15 years in the global telecommunications sector has been the privatization of state-owned and operated assets and the opening of markets through liberalization. The reasoning for why this agenda has been accepted and implemented throughout the world as a strategy for telecommunications development probably comes down to a simple equation. New telecommunications technology requires money, and privatization brings in money. So, especially for developing countries hungry for capital and investment, privatization is an agenda to which both corporate and government institutions can agree.

But the values of liberalization are not as clearly defined and are, as one may expect, much more a point of contention. Because there are so many corporate institutions around that want to play the game and build telecommunications infrastructures, the market should allow them to flourish—within manageable bounds, of course, to ensure that certain social needs are met. But what is the criterion for the common management of telecommunications development in a competitive environment? Until those values are defined and agreed upon by corporate and government institutions, the debate will remain and the contention will hinder our ability to connect telecommunications development with economic growth, social modernization, and political participation.

It is at this point that we enter the debate in earnest and begin to sketch out the possible responses to the challenges of institutional transformation. It is important to begin, then, with a discussion of the values that underlie the strategies of privatization and liberalization that have been at the foundation of recent telecommunications development programs. From there, various kinds of liberalization schemes can be examined and compared, thereby launching the broader discussion of strategic liberalization as a policy proposition and as the basis for corporate strategy.

Ownership

Even though public and private-sector institutions are very different, there are a number of similarities when it comes to the problems and techniques of management. A politician, as a representative of the people, chooses to use the authority available to change policy on a given issue. A corporate chief chooses to implement one plan over another, and alters the balance sheet and the kinds of products and services available to the community the corporation serves. Both of them are working to control resources in a fashion that serves either their interests or the interest of the community as a whole.

Nevertheless, it is the different ownership structure of private and public institutions that determines many of the differences between the two. Ownership determines the purpose of management by articulating the business goals of the company. In turn, those goals shape the character of the job functions and the measurements for success and failure. In the case of economic institutions, such as a company that provides telecommunications services, one can reasonably speak of a continuum of ownership options available: total ownership by the state, to total ownership by the private sector. There are various alternative arrangements that exist along that continuum, and many economies in transition have decidedly complex arrangements whereby ownership between government and the private sector is shared.

The main difference between the two extreme cases is the management goals. The goal of any privately held corporation is ostensibly to maximize the value of shareholder investment (O'Reilly, Hirsh, Defliese and Jaenicke, 1993). It is the shareholders who determine the management by appointing directors, who then appoint executive officers to carry out the management directives. Publicly held companies are often in the position of having their goals set for them by the politicians. Returns on investment are tightly regulated, competition is simulated through certain kinds of incentives and disincentives to invest, investment requirements are mandated, and personnel is often assigned by the political decision makers.

Differences between public and private-sector institutions are defined, in great part, by this fundamental difference in goals. Even though the elements of bureaucratization, a professional managerial strata, and technological implementation may be similar in a public and private-sector institution, the fundamental orientation of the institution is nonetheless different. The responses to the external environment are also different; a private-sector institution relies only on its ability to respond to competitive change, while a public institution often concentrates on the single shareholder that has a vested interest in its success and profitability (namely, the government).

This difference in orientation is especially important with regard to the provision of infrastructure services, which have often been defined as "strategic" sectors for economic development. Roads, bridges, telecommunications lines, water, and energy are critical components of the development process, and represent not just economic capital for citizens, but also political capital for government officials. For many of the modern services essential to the development process the traditional kind of management was a centralized state bureaucracy that allocated all resources and made all decisions. The transportation ministry took care of the roads. The ministry of communications took care of the phones. For many countries, this arrangement weakened these institutions, which often could not be separated from the political institutions that managed them (Duch, 1991; Petrazzini, 1995). This, in turn, weakened the ability of states to use public enterprises to contribute to the goals of development (Clapham, 1985).

As a result, public and private-sector decision makers have turned to privatization in an attempt to make these public sector institutions more responsive to the interests and needs of citizens.[2] The reasons suggested for this trend are more than enough to fill an entire book but, in the end, it comes down to performance. In this circumstance, performance is defined as the ability to sustainably increase the capacity and capability of the institutions providing services to citizens. Improving performance becomes the key to the discussion, in both the developing and developed world.

In order to achieve higher levels of performance, privatizations of formerly monopolized public enterprises have taken place throughout the globe. Starting in Great Britain during the Thatcher administration, policy makers found a new opportunity to turn over the management and provision of certain services to the private sector. Although Great Britain did not privatize more than 5% of existing social and government services, the big headlines garnered by a few of them were large enough to be seen around the globe (Redwood, 1987). British Telecommunications completed the £ 20 billion sale of shares to the public in 1993, after 7 years of preparation and initial sales.

In the United States, a different kind of market shift was occurring. The negotiation of the 1956 consent decree marked a redefinition of the activities of AT&T as the regulated private monopoly provider of telecommunications services in the United States and increased the pressure for further liberalization and deregulation in the competitive marketplace. The confrontation between the Justice Department, with its historic role as the enforcer of antitrust legislation, and AT&T reached a new modus vivendi in 1982, when a modification of the 1956 decree was proposed. The agreement broke up the Bell System, spinning off seven Regional Bell Operating Companies (RBOCs) to provide local telephone service while reconstituting AT&T as a long-distance provider and manufacturer of telecommunications equipment. The centralized model for telecommunications development collapsed under the weight of political pressure, opening the door for a new era of competition in the United States.

The fall of the Berlin wall served to further discredit the centralized model of state-sponsored institutional development. Communist governments that had once owned all the means of production according to good Marxist-Leninist traditions have sold everything from the big Stalinist steelworks right down to the corner kiosks. Businesses have been sold to the private sector, sometimes distributed to citizens in the form of privatization vouchers, sometimes auctioned, and at times simply given away to those who had the ability to manage the assets (Frydman and Rapaczynski, 1994). In both East and West, the privatization decision came down to performance: The ability of economic institutions to sustain growth and profitability depended on reducing the linkage between the public and private sectors.

At the same time, the developing world faced economic pressures that made privatization an even more attractive policy. Most countries in the developing world found themselves strapped for cash after the drop in oil prices at the end of 1982. With overextended lines of credit, countries began to face the realities of difficult austerity programs and severe cutbacks of social services. Increasingly, the privatization of traditional state-sponsored services began to look more and more attractive as an opportunity to receive fresh and productive capital investments. The trend has spread to even the poorest countries of the world, where many governments have begun to consider what privatization can do for their economic and political interests. The issue, again, is the performance of these institutions, ensuring their ability to serve the people who are customers and recipients of critical services.

Privatization is meant to improve performance in three areas:

• Reducing the interference from government officials, thereby permitting a more efficient allocation of productive resources;

• Changing the property rights that define the administration of the enterprise, diminishing the information gap between owner and administrator and defining appropriate incentives for management; and,

• Removing the possibility of government subsidization, which improves financial discipline (Jones, Tandon, and Vogelsang, 1990).

Privatization is therefore meant to have a macroeconomic and a microeconomic impact. The macroeconomic impacts are increases in foreign exchange if foreign investors take part in the privatization process, a reduction of public debt (which is often part of a privatization package), and a better fiscal picture for the country as a whole. Privatization is meant to restructure the operations of the company so as to ensure the improved performance sought after by each of these countries.

It is also possible to speak of a macro and micro political impact. A privatization policy removes from the hands of government one of the strongest potential levers to garner political support: political patronage through jobs in state-run enterprises. The long-term dynamics of the political system will be altered, especially in countries whose dominant part has traditionally been connected to labor unions. That means systemic political change and the changes for the individual institutions that embody political authority.

The immediate result of this privatization trend has been a reordering of capital flows worldwide; from the growth of diversified mutual funds in the developed countries to the increased lending from major international institutions, the developing world and the former east bloc have been soaking up capital in this wave of privatizations. But, on the microeconomic and micropolitical level, there is still only incremental change. Many of the now privatized enterprises throughout the world still have a long revolution to go before the management of those enterprises is truly transformed. There is one simple reason for the limited degree of institutional change on this level: It is not privatization which compels alterations in institutional structures, but competition.

Competition

Ownership is not the only determinant of management and institutional structure. If it were, privatization would be the end of the story. We would find universal characteristics of all publicly and privately held companies, determine which was better, and move on (Duch, 1991; Petrazzini, 1995).[3] Ownership sets ostensive goals and a framework for management, but it is the response to the environment external to the corporate institution that determines economic behaviors and institutional practices. Characterizing the environment, therefore, is almost more important than characterizing the ownership structure.

To illustrate this point, we take the polarities of public and private sector management and overlay another dimension: Competition and monopoly. Competition, on one hand, is traditionally defined by economists as an environment where a number of institutions offering similar or identical goods or services have access to a customer and compete for a customer's money. Monopoly is a condition where one firm supplies everything for a particular product or service within a specific market. Although there are many different kinds of monopolies and classifications for competitive environments offered by economists, this portion of the discussion defines competition generally so we can focus on the differences between monopolies and competitive institutions.

Without delving into the complex political and economic justifications for monopolistic versus competitive marketplace arrangements, it suffices to say that the management of a competitive enterprise and the management of a monopolistic enterprise are completely different. The production of a monopolist is determined not by marginal cost but rather by economies of scale. Because the addition of other producers would increase the overall cost of production, economists have argued that certain industries have the characteristics of a natural monopoly (Baumol, 1988).[4]

The alternative environment to a monopoly is competition. Generally put, a competitive environment has a number of different service providers that attempt to gain market share for their products through product differentiation based on price or quality. Classic microeconomic theory has a number of definitions for different kinds of competitive markets, from oligopolistic competition with its constraints to market entry to a open market competition where there is no barrier to entry and real price differentiation.

It is important to note that a large number of firms do not necessarily mean that there is a competitive market. Likewise, the absence of more than a few players does not indicate that an oligopoly exists (Stigler, 1968; Toulan, 1994).[5] The critical issues are the barriers to entry and the ability of entering institutions to sustain themselves in the face of aggressive pricing strategies on the part of existing players. If because of regulation of technological cost, the barriers to entry are too high, or, if for reasons of regulation or preponderant market advantage, competing institutions can not sustain themselves, the market for a particular service can not be characterized as competitive.

Although changes in ownership structure have occurred in many places throughout the world, we have seen less of an attempt on the part of governments to create a competitive environment for the provision of infrastructure services. In many cases, it can be argued that privatization has left in place a private monopoly where once a public monopoly existed. "If the government wants privatized enterprises to play an efficient social role," one commentator notes, "the regulatory regime should foster a maximum level of competition" (Sanches and Corona, 1993).

This is not to say that privatization represents a less than significant change but, rather, it is to point out that privatization is not and should not be the whole story. Far too often, political and economic commentators link privatization to free-market economics without recognizing the fact that a free market for goods and services requires, by definition, open entry for other competitive concerns; a privately held but tightly regulated monopoly certainly does not allow for such a condition. Competition requires the wholesale alteration of certain political and economic institutions that determine the provision of services under a monopoly regime, a shift that is more of a transformation than a shift in ownership structure.

A recent study by the Organization for Cooperation and Economic Development (OCED) on the impact of privatization offers the following perspective on increasing competition:

> Market competition lies at the root of economic efficiency. The transfer of ownership rights results in improvements in efficiency, but these are limited by the uncertain effect of privatizations on market structures. In the operations carried out in the sample countries, we are bound to note that the transfer of ownership has not particularly encouraged the development of competition. In certain countries, privatization has even strengthened the movement towards industrial and financial concentration (see the examples from Chile from 1973 to 1982 and Mexico since 1983). From the viewpoint of the impact on economic efficiency, the most satisfactory method is to increase the role of market forces during the privatization process, in order to avoid the exploitation of oligopolistic or monopolistic market structures for private ends. (Bouin and Cichalet, 1991, p. 14)

In other words, liberalization establishes competition by reducing and eliminating restrictions which protect monopoly service provision. That means less (or at least different) regulation, prices, and quality of service demands that respond to market forces, and new entrants into the market.

What are the benefits of a competitive transformation? This book argues that if government and corporate institutions are to act together to push down the cost of essential infrastructure services and generate worldwide technological innovation, it will be necessary to introduce further competition into markets that have been heretofore monopolized. By pushing down prices and setting the stage for introducing new technology, essential goals of national development will be realized.

Fig 1.1. The path of institutional transformation in the telecommunications industry.

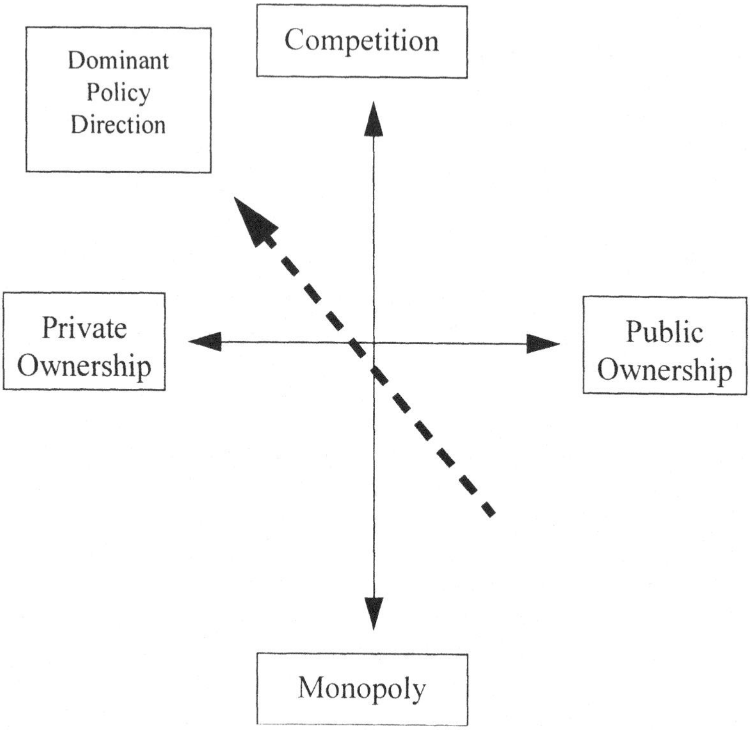

```
                          ┌──────────────────┐
┌──────────────────┐      │   Competition    │
│    Dominant      │      └──────────────────┘
│    Policy        │               ↑
│    Direction     │               │
└──────────────────┘               │
                          ↖        │
                            ╲       │
┌──────────────────┐         ╲      │        ┌──────────────────┐
│    Private       │←─────────╲─────┼───────→│     Public       │
│    Ownership     │           ╲    │        │    Ownership     │
└──────────────────┘            ╲   │        └──────────────────┘
                                 ╲  │
                                  ╲ │
                                   ╲↓
                          ┌──────────────────┐
                          │     Monopoly     │
                          └──────────────────┘
```

In the words of the 1994 *World Development Report*:

> While the special technical and economic characteristics of infrastructure give government an essential role in its provision, dominant and pervasive intervention by governments has in many cases failed to promote efficient or responsive delivery of services. Recent changes in thinking and technology have revealed increased scope for commercial principles in infrastructure provision. These offer new ways to harness market forces even where typical competition would fail, and they bring the infrastructure user's perspective to the forefront. (World Bank, 1994, p. 13)

Liberalization is the key to realizing the benefits of privatization. In turn, it is an essential part of any strategy to link the benefits of infrastructure investment directly to the needs of people.

For the past century, monopoly has been the assumed fact of economic production and political organization for these industries. But technological, social, and economic changes have altered the environment for many of these industries. The dominant trend in the institutional transformation of the telecommunications sector, graphically depicted in Figure 1.1, has been towards increased competition and private ownership of telecommunications networks. There is now an opportunity to reassess our understanding of these institutions and change the environment in which these institutions act.

The problem is, that is where the consensus breaks down. Although most commentators largely agree that liberalization is an appropriate response to institutional change, there is a wide range of disagreement on what kinds of liberalization are appropriate. Again, this is in great part because there are few shared common goals that have been articulated and supported by the public and private sector institutions driving the course of the transformation of the telecommunications sector; a variety of economic, political, and social justifications are brought to the fore in the analysis of the future of the telecommunications sector. It is now time to look more closely at some of those paths to liberalization and how they define shared values for action.

PATHS TO LIBERALIZATION: STRATEGIES FOR TELECOMMUNICATIONS DEVELOPMENT

Figure 1.2 lays out some of the most prevalent theories and perspectives on the future direction for the global telecommunications sector, and offers a basic classification of some of the policies suggested for the transition from a market dominated by state-run monopolies. This is certainly the dominant direction in international telecommunications policy today, and the need for increased competition and infrastructure investment is largely agreed on by the broad spectrum of analysts and commentators. What the figure makes clear, though, is that there is still a wide range of opinion on what principles should guide the evolution of national and global telecommunications infrastructures.

Cultural and Technological Protection

On the left are those who subscribe to the theories of cultural and technological protectionism. The concern for these advocates is the centralization of media and telecommunications investment throughout the world and the increasingly large gaps between the technological "haves" and "have nots." Writers such as Jill Hills, Gerald Sussman, and John Lent have drawn heavily from the tradition of dependency theory to advocate a central role for the state in protecting the cultural heritage of various national groups, as well as the technological capabilities of local firms (Hills, 1986; Sussman and Lent, 1991).

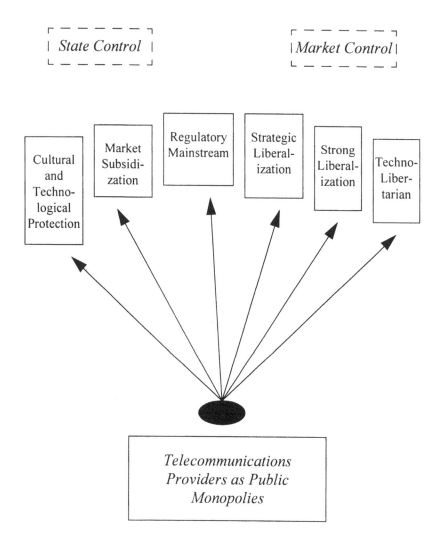

Fig 1.2.
The paths to liberalization: Theoretical models.

The underlying rationale for this position is that companies, especially multinational ones operating in the developing world, maintain business practices that are diametrically opposed to certain aspects of the development process. For the writers associated with this school, the telecommunications sector is perhaps one of the most significant culprits in the erosion of local culture, language, and self-esteem among ethnic minorities in developed countries and the entire populations of developing countries. Television shows are dumped from the United States at cut-rate prices, undermining the viability of local sources of information and entertainment content. "Cherry picking" and "cream skimming" damage the telecommunications infrastructure, which is oriented not to serve the needs of the people in the developing world, but rather the needs of the corporations that are operating across national boundaries. People become "dependent" on what Hollywood has created for entertainment and cultural affiliation, and accept, without recourse, the infrastructures that are built to support the services for which only the wealthy can pay.

The institution responsible for addressing the difficulty is the state, which has at its disposal a variety of effective tools to counter the encroachment of the private sector on the needs of the people. Quite often, writers from this school call for the maintenance of quotas on foreign films and entertainment for local broadcasting, high percentages of local ownership for all telecommunications and entertainment assets, and strict guidelines for the operation of multinational companies within their borders.

One of the success stories which many writers from this school point to with regard to the telecommunications sector is Brazil's efforts in building an indigenous market for equipment, such as digital switches and the like. The import substitution strategy carried out by the military dictatorships of the 1970s and 1980s lead to solid growth in the manufacturing portion of the sector; but, at the same time, the dependency theorists point out that this infrastructure largely served the economic and political purposes of the oppressive government establishment.

Market Subsidization

Closer to the center are policy researchers and commentators like Eli Noam and politicians like United States Vice President Albert Gore. They focus on many of the same concerns as the advocates of cultural and technological protection, but are more likely to suggest subsidization and the imposition of certain social goals, such as universal service, on information and telecommunications providers (Noam, 1987, 1994). The economic and social models that bolster this kind of thinking include the traditional analysis of scarcity in the telecommunications industry, and the belief that the sector can not sustain widespread facilities-based competition.

Although Gore's position has changed somewhat since taking the trip down Pennsylvania Avenue from the Senate to the White House offices, his policy statements in the early 1990s still reflect what is considered by many to be an appropriate direction for corporate and public policy. In an article in *Scientific American* in 1991, he wrote about the need for public commitment to infrastructure development in order to counterbalance the effects of competition. That policy was embodied in his proposal for the National Research and Education Network (NREN), passed through Congress and was signed into law while George Bush was still President. It a reflection of how "market subsidization" might work in practice.

NREN represented a multiyear, multibillion dollar commitment on the part of the U.S. government to build a network specifically to meet the needs of primary and secondary schools throughout the country. The view was that these schools would not have the resources necessary to invest in the technology alone, and a nationally funded and sponsored initiative would provide the scale and scope necessary to ensure needs were met. So the model for "market subsidization" is pretty straightforward: The government looks at the evolving market, defines a need specific to the common good, and invests to ensure that a certain kind of access and service is provided.

The Regulatory Mainstream

Then there is the regulatory mainstream, typified by the work of Michael Tyler and the predominant official policies of the World Bank (Saunders, Warford, and Wellenius et al., 1994; Tyler and Bednarczyk, 1993; World Bank, 1994). In terms of telecommunications development, the goal of those in the regulatory mainstream is to identify services that are most appropriately competitive, liberalize them individually over an extended period of time, and maintain a close watch over the marketplace through traditional regulatory institutions. The focus is on defining the "best practices" of regulatory institutions and ensuring that there is some commonality throughout the world in terms of approaches to telecommunications regulation.

The most notable work on how telecommunications development can be promoted through a regulatory mainstream comes from Alan Saunders, Jeremy Warford and Bjorn Wellenius at the World Bank, whose book, *Telecommunications and Economic Development*, is thought by many to be the key text in the field. The models presented in the book, many of them still from the 1970s and 1980s even in the most recent edition printed in 1994, focus on a case-by-case approach for the justification of competition in certain markets, the need for regulation and restriction in others, and the successes of multilateral funding in the telecommunications sector.

In these kinds of writings, the state is still the more important locus of activity, so perhaps the arrow in Figure 1.2 should be tilted a bit to the left rather than straight into the center between state oriented and market oriented approaches to liberalization. Nevertheless, this piecemeal approach often allows for a greater degree of market oriented liberalization because of the economic focuses of the methodology, rather than the political economy of writers like Noam and Gore.

Strategic Liberalization

In Figure 1.2, there are three policies listed which try to move the focus of the competitive transformation away from the state and regulatory institutions and toward the institutions responsible for providing the services, namely the telecommunications and information companies. The next policy listed on the center right is Strategic Liberalization. Building on the work of Russ Neuman of the Fletcher School of Law and Diplomacy and Lee McKnight of the Massachusetts Institute of Technology, strategic liberalization argues for the implementation of specific policy measures to increase competition in the market for wireless access services such as cellular, wireless local loop and satellite communications. This point of view argues that a strategic focus on investment in and the development of wireless communications will provide a sustainable foundation for the continued growth of the telecommunications sector and the successful introduction of new products and services in an environment of facilities-based competition. But before speaking further about strategic liberalization and the factors which differentiate it from other theories in the field, there are remaining boxes on the right discuss.

Strong Liberalization

Strong liberalization is a term used to describe the work of Greg Staple and Peter Smith, both of whom have argued for a more complete liberalization of the telecommunications industry than is advocated by those in the regulatory mainstream or on the left (Staple and Smith, 1994). Their argument is centered on the proposition that liberalization in niche services, such as data transmission and value-added services, should take place in conjunction with the introduction of competition for all services in the local loop.

Staple and Smith's work on chronicling the liberalization efforts through Asia provide the foundation for their argument that introducing competition at all levels is an appropriate model for telecommunications development. Many of the success stories they point to involve broader forms of liberalization than those which would be advocated by the thinkers in the political mainstream. Their basic assumption about the evolving nature of the industry is different as well; the strong liberalization approach points to the successes of countries like New Zealand and suggests that the telecommunications sector should not be treated any differently than other sectors of economic activity, at least from the perspective of regulation and public policy.

Techno-Libertarians

Finally, there are techno-libertarians like George Gilder and Peter Huber, who argue from the basis of technological opportunity (Huber, 1993). The techno-libertarians feel that the technologies of the emerging telecommunications marketplace

are inherently competitive and, if given the freedom to grow and expand, will provide a sufficient foundation for sustainable market conditions. The "libertarian"
ethos inherent in the writing of Huber and Gilder is reflected in their strong belief
that the technologies of the telecommunications revolution are inherently competitive; they do not promote market hegemony, and provide more than sufficient opportunities for direct, facilities-based competition between telecommunications
companies and information service providers.

That kind of approach leads to arguments that many consider radical. For example, in the lead-up to the auction of new licenses for Personal Communications
Services in the United States during the last year, Gilder took the position that
emerging technologies in the wireless arena would alter the environment to such
an extent that licenses would no longer be need to be assigned to telecommunications providers at all. Huber, as of mid-1997, was working on a book arguing for
the outright abolition of the Federal Communications Commission in the U.S.

Comparing Values for Liberalization

There are pieces of all these approaches embodied in the corporate strategies and
government regulations in the telecommunications and information sectors
throughout the world. For example, there are many governments that demand certain percentages of locally made television shows and that have resisted any foreign ownership of public telecommunications infrastructures, quite in line with
the analysis provided by the school of technological and cultural protection, while
at the same time passing laws which stipulate the need for liberalization and increased competition in the telecommunications sector, quite often using the justifications espoused by the techno-libertarians.

So much for consistency in the minds of government officials. But, in many
ways, corporations have been equally as inconsistent in terms of their strategies.
Institutionally, most telecommunications providers are still the owners and operators of the public network and act as if they are government owned, even if they
have been privatized and operate in a nominally competitive market. At the same
time, their public relations machines turn out statements about how important
competition is and how new technologies will unlock the energies of the marketplace.

What it comes down to is values. Fortunately, because this book is being written about telecommunications and not about Presidential politics in the United
States, talking about values should not lend itself to difficult and irreconcilable po-
lemic. In a time when the transformation of the telecommunications sector has
been identified as one, if not the, critical driver of growth and opportunity in the
next century, a healthy discussion of values, not just policies, is important. The
values outlined in the telecommunications development strategies mapped out in
some detail before, though, do not match the needs of the times.

Take, for example, the value of "personal freedom" often associated with telecommunications technology. The value of freedom expressed on the right and

left sides of the spectrum are certainly in opposition. The perception among the techno-libertarians is that freedom comes from the technology, while those who advocate cultural and technological protection point out that freedom can be taken away by the technology. Or, in the center, freedom is somehow modified by various levels of government control of competition or facilitation of sector evolution. The language can be bent to the point of breakage, not leaving us a shred of connection to draw the sides together.

So what of strategic liberalization and its values? Is there enough substance in them to provide for a common direction to motivate corporate and government players in building an appropriate infrastructure for competition and telecommunications development in the next century?

THE THEORY OF STRATEGIC LIBERALIZATION: IDENTIFYING NEW OPPORTUNITIES FOR DEVELOPMENT

Strategic liberalization differentiates itself from all of those approaches in two critical ways. First, it is based on the proposition that the history and development of certain technologies offer greater opportunity for the development of sustainable, facilities-based competition than others. Such a fact opens the door for a constrained, targeted contribution for the state in facilitating the transformation of the telecommunications industry.

Second, and perhaps most importantly, the perspective of strategic liberalization incorporates the perspectives of mainstream political and social scientists who are particularly interested in the problems of development, but who do not subscribe to the arguments of dependency theorists and protectionists on the left portion of the political spectrum. The rich tradition of comparative politics forms the basis for many of the values expressed in the policy of strategic liberalization.

Certain kinds of competition are more likely than others to connect telecommunications development to the traditional goals of national development, such as reductions in the incidence of poverty and increases in the resources made available to the members of each nation's citizens. To achieve those goals, there is a great wealth of literature in the fields of economic and political development to draw theoretical and practical direction. Identifying specific changes within the technological, economic, and social fabric that present new opportunities for development will set the groundwork for discussing an appropriate policy implementation program for strategic liberalization.

Telecommunications Development and National Development: Setting the Groundwork

Development is a messy process. No matter what the formal definition, no matter what the benefits and costs that technological and social advances may have for a community, broad social change is not a simple thing for those who have to live through it. Even when that social change comes in the form of nominally defined technological and social advances, altering the structures on which our present lives depend creates an uncertain future. And no one likes too much uncertainty.

For that reason, there are as many different definitions of development as there are kinds of social and economic uncertainty. There is the kind of development that focuses on jobs and higher wages, a reaction to the fear and reality of unemployment and underemployment. There is the kind of development that concentrates on open political participation, which counters the fear of disenfranchised economic, ethnic, or racial minorities. There is development based on technological revolution, founded on the demand for progress rooted in all kinds of social fears and concerns.

Students of the literature of development employ common descriptions of the development process: Countries start low, move up the development ladder, and begin to take on the characteristics of the developed world. Or, to use language that has become largely obsolete after the end of the cold war, countries move from the third world (or the fourth world), into or past the second, then up to the first world. Institutional needs and characteristics change as development proceeds. The strategic management of resources attempted by corporate and public institutions during the process alters as new institutions arise to stake out their claims of authority over economic sectors, geographical regions, or communities of interest.

The textbook definitions bear these distinctions out, but also add a fundamental principle which will be critical to our ongoing discussion. One of the most widely available textbooks on developmental economics refers to development as a fundamental change in the economic and political structure of a community such that the individuals in the community become the major participants in the process that brought about the change (Gillis, Perkins, Roemer, and Snodgrass, 1987). The implications of such a definition are clear enough: Management of strategic resources, such as infrastructure or natural wealth, needs to be enhanced and focused by making members of the community more productive and by improving the quality and character of leadership. The value to be enhanced through policy and research is the enfranchisement of citizens through changes in the economic and political process.

This is where serious analysis of institutions becomes critical in the overall discussion of developmental theory and practice: The management of strategic resources is guided by the corporate and government institutions that define the development process. These institutions act as a center of gravity to attract social meaning and energy, using the bureaucratic science of the modern age to provide a rational and legal structure to manage participation by members of the community (Weber, 1948).

Under the pressure of evolving technology and shifting social demands, institutions are altered. For example, agricultural institutions changed dramatically at the time of the industrial revolution, and labor demand shifted from the countryside to the cities. This is nothing new. The agricultural revolution that occurred between the 3rd and 4th millennia BC radically changed communities from hunter gatherers to cultivators of crops and agricultural surpluses (Renfrew, 1990). The organization of society changed with the technology of agriculture, altering both the language and physical forms of communication between peoples.

But just because social, economic, and political change is nothing new does not mean that it is any better understood. That is partly because the character of the changes differ from age to age. If we are to take a basic taxonomy of institutional change over the last 500 years, for example, we would find a wide variety of social and economic institutions rising and falling without finding a fundamental explanation that would apply to all cases.

That is in great part because the institutions that support certain social and economic practices vary from community to community and, within each community, certain kinds of social configurations are more important than others. A prevalent example in today's discussions on privatization throughout the world concerns the institution of the credit market in certain European countries as compared to the United States. German banks traditionally have wide latitude in determining the management of those companies where their money is invested, while in the United States, banks and lending institutions do not play as central a role in the private-sector management of other economic institutions. That historical arrangement has a definite impact in a privatization program, for example, by defining how financial instruments will be structured and how people will understand the change in economic and social relationships.

Differences in institutional arrangements are reflected in a wide variety of social facts, although the linkages between institutions and facts are not as tangible as many social scientists would like. The very idea of credit, for example, depends on a sense of mutual trust. A lending institution believes that you will pay the money back after they have loaned it to you, and has taken steps to assess whether or not you are telling the truth about your ability to repay. What that trust means, and the character of that trust, is at the core of many arguments about development.[6]

One thing we can be sure of, though, is that social facts such as trust do have a direct impact on institutional performance, even though they might not easily be quantified or accounted for in the courtroom, legislatures, or boardrooms of the world. Social facts and values determine the success or failure of business ventures, the sustainabiliy of investment and growth and, to a significant degree, the success or failure of efforts to achieve political and economic development.

There are many possible ways to begin addressing the relationship between institutions, values and social facts. For our purposes, though, the best starting point is offered by Robert Putnam. He grounded his work on civic traditions in Italy with one basic question: "What are the conditions for creating strong, responsive, effective representative institutions?" (Putnam, 1993). It is the issue of institutional performance that will be critical for us as we assess the impact of competition on the public and private institutions of the telecommunications sector.

Putnam began with two fundamental points:

- *Institutions shape politics.* The roles and standard operating procedures that make up institutions leave their imprint on political outcomes by structuring political behaviour. Outcomes are not simply reducible to the billiard-ball interaction of individuals nor to the intersection of broad social forces. Institutions influence outcomes because they shape actors' identities, power, and strategies.

- *Institutions are shaped by history.* Whatever other factors may affect their form, institutions have inertia and "robustness." They therefore embody historical trajectories and turning points. History matters because it is "path dependent": Whatever comes first (even if it was in some sense "accidental") conditions what comes later. Individuals may "choose" their institutions, but they do not choose them under circumstances of their own making, and their choices in turn influence the rules within which their successors choose. (Putnam, 193, p. 7)

Clearly, if we want to change the world in which we live, we need to construct institutions that take root in the social environment in which they are planted. Putnam's guidelines tell us a few things about how to judge the sustainability and effectiveness of institutions. First, no institution lives in a social vacuum, which means they can not be engineered in a vacuum. There is always an element of social engineering at the basis of public policy, insofar as the very idea of policy assumes that institutions shape politics, but the tendency is to start as if the slate were clean and institutions can be constructed as if in an ideal world. Putnam made clear the impossibility of such a project, which means that our search for sustainable institutions that can serve the objectives of national development has to be infused with historical and social facts.

Second, analysis should focus on how institutions interact with each other, and determine some of the patterns and dynamics revealed by institutional interaction. Private and public sector institutions, for example, establish historical trends that shape the viability of emerging institutions and the possibilities for the transformation of existing institutions. A closer examination of the patterns of institutional interaction should tell a great deal about what kinds of developmental needs require evolutionary institutional change, and which ones necessitate revolutionary change.

Connecting Institutional Performance to Telecommunications

In the last section, we used the example of the credit markets to illustrate a number of points about developmental theory. Certainly, credit markets have been critical for the developed world's progress from the Enlightenment all the way up to the present day. But credit markets are far from being the only source of institutional energy to spur economic and political development. Other kinds of institutions have moved to the forefront in our discussions of modernity and development in the past 100 years. Of critical concern in the modern age are the institutions that provide the infrastructure that is required of developed societies: power, roads and communications.

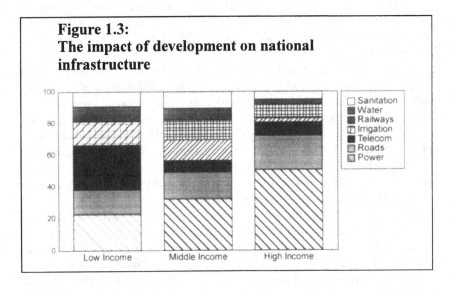

Figure 1.3:
The impact of development on national infrastructure

Figure 1.3 is adapted from the *1994 World Bank Development Report*, and provides a basic overview of the institutional change that would likely occur as a community "develops" from a low-income to a high-income community. The graphic shows one of the most significant elements of this institutional shift: Countries with higher per capita income levels are likely to have a different mix of infrastructure than those with lower levels of per capita income. The process of development, as most commonly perceived, is a linear trend directed by the arrow of time; countries move from lower to higher income levels as they develop resources. In the process, different economic institutions become more significant.

Figure 1.3 clearly shows how the economic infrastructure of a developing country changes in the process of development. Of those infrastructural elements, three become more important as income levels increase: power, roads, and telecommunications. Most would argue that power, roads, and telecommunications have brought improvements in the quality of life to those individuals who have access to the services. Power has allowed us to function more effectively and to produce more. Roads have given us further mobility. Telecommunications have given us new sources of entertainment, information, and education. The application of these technologies has contributed to increased literacy, improved standards of living in terms of per capita product and income, decreased incidence of disease and infant mortality, and a host of other changes considered real improvements in the quality of life worldwide.

Change in the infrastructural mix is an adaptation to those social and economic needs, and, in turn, changes in infrastructure reinforce social and economic trends (Milliken and Blackmer, 1961). In developed countries, there is a highly mobile work force and capital base, which depends on and often contributes to the construction of infrastructure resources. The transition from a manufacturing to a service economy that has been a consistent theme of economic and political literature in the developed world has depended, in great part, on the technologies of the modern (or, even, postmodern) age.

These services are critical to the developmental process, and have been the subject of special focus and concern by researchers and decision makers in this field for many years. The most common connections between these "modern" services and development are drawn through variables such as productivity, social and economic mobility, and quality of life factors such as the health and welfare of all members of society. The particular analysis of telecommunications investment on economic and political development has covered a great deal of ground in linking sector development to all of these factors, and some of that research will be reviewed in more depth later in the chapter.

For the moment, the connection that has to be drawn is between the idea of institutional performance and the specific challenges inherent in managing these modern infrastructure services. Because of their critical importance to the overall growth and development of a country, they require special attention from policy makers and corporate managers—there are social implications to telecommunications development, even in the most competitive of markets.

Second, given the dimensions of institutional performance previously mapped out, an active but appropriate role needs to be defined for policy that goes beyond establishing a framework for competition, especially for those countries in low income categories. By promoting certain values in the development process, resources can be oriented to specific needs and opportunities that development makes possible.

Appropriate Values for Telecommunications Development

But what values are most appropriate for policy in the telecommunications sector, given that framework? What kinds of institutions can carry and promote those values within the context of continued telecommunications development? Three values in particular will be critical to the global success of telecommunications development:

- Political and economic participation;

- Social and technological modernization; and,

- Creating effective public policy.

The discussion of how prevalent theories of economic and political development address these values will lead us back to a set of propositions about constructing a new institutional environment to foster competition in the provision of certain services.

Political and Economic Participation

Any proponent of any of the liberalization strategies listed would agree that a significant opportunity exists to define new channels for economic and political participation through emerging technologies. The literature of development makes it clear that better participation means better production by firms and more responsive political institutions—these are goals in and of themselves for the development process. If liberalization can focus on providing a further impetus for economic and political participation, it will better contribute to the goals of national development.

It is a fundamental assumption of the modern age that a society is only productive when it uses the capabilities of each of its members to the fullest. This is true in both an economic and a political sense. Individuals who do not participate in the political system do not do as much as possible to promote its vitality, and without educated citizens, participatory institutions are likely to falter. Those who consume more in services than they are able to give take away from the economic efficiency and productivity of a nation, and those who opt out of the economic system do not promote the level of allocative efficiency that a market structure has to offer (Hirshman, 1970). By allowing the open marketplace to aggregate the needs and wants of a society, many thinkers presume that the direct connection between services and people in both the political and economic spheres is best made.

Development theory is "ultimately concerned with what people can or can not do" (Sen, 1970). But the how and what of this statement depends largely on the issues that the individual brings to the table. For Amartya Sen, the author of the quote just presented, a person's action is determined largely by his or her entitlements, defined as the commodity bundles a person can handle. Although the definitions and terminology are very different, a political scientist like Barrington Moore would express the same concern: Political participation is based on the kinds of opportunities offered to the citizen. The viability and success of the system depend on the opportunities available.

How we participate, both economically and politically, is radically changing with the introduction of new technology. It is technology that defines the opportunities for participation in any open society (Innis, 1954).[7] For example, the opportunities to receive information through televisions and radios have dramatically altered the political landscape in developing and developed countries (Katz, 1977). The possibility of electronic plebiscites and direct democracy in the developed countries of the world has captured the imaginations of some who see information technology as the great liberator of democracy (Barber, 1984).

In the last 20 years, we have seen a dramatic shift in the technologies of communication, from the computer and digital revolution to the possibilities of wireless communication that are discussed in this book. These new technologies are emphasizing horizontal relationships over vertical ones, connecting individuals to each other and reordering corporate and public institutions (Neuman, 1991).

Many have spoken of a "flattening" that needs to happen among large, bureaucratic institutions so that they become more responsive to the needs of the people they employ and serve (Goldman, Nagel, and Preiss., 1995). That kind of institutional shift has broad consequences for the opportunities for participation in the process of development. Instead of concentrating solely on vertically oriented management of resources, we can speak again of horizontal relationships as the foundation for resource allocation and open political communications.

As institutions flatten and move to a new structure of internal and external communication, they will have an opportunity to improve both the quality and quantity of participation. Because development depends on the ability of social and economic organization to enhance the productive and effective participation of its people, changes in the structure of participation represent a unique opportunity and challenge.

But, in order for people to use technology to participate in improving the economic and political institutions, they need to have access to that technology. A common concern voiced by many writers is the very real possibility that there will be a technological divide between rich and poor.[8] There is already a clear distinction within and between countries when it comes to technology use; the poorer the country or community, the less technology like telephones and electricity is used (Egan and Wildman, 1994).

In a world of scale economies, the easy answer has been to subsidize the users who could not afford the service. Combined with further technological penetration through education, subsidization policy is certainly a critical element in enhancing economic and political participation through new technological opportunities. In a rich country or community that wishes to make that political decision, it is possible to subsidize to the point where we can reasonably speak of 100% penetration by a service. But few countries have that luxury, and even communities within the richest country in the world do not have that luxury. The only starting point for addressing this problem is to lower the cost of service as much as possible so that it can be made available to as many people as possible.

When we look at the opportunities for liberalization in light of the changes in the possibilities for participation, we clearly need to concentrate on low-cost technological applications. If liberalization only touches those areas where high-cost technology and services are dominant, we will miss the opportunity to use liberalization as a tool to enhance participation and, through participation, development.

Wireless technology has distinguished itself as a potentially low-cost solution to the issues of universal service and access, in both the developing and developed worlds. At present, costs for wireless access are less distance sensitive, thereby offering access solutions to rural communities and geographically isolated regions (Hudson, 1984, 1994). But it is clear that the costs of wireless access services, when compared to wireline access, will soon be significantly cheaper, even for residential and business communities in highly populated cities. This alteration of the cost curves for wireless and wireline access services represents a significant opportunity to employ liberalization to reinforce a significant trend: decreasing costs for access in the telecommunications industry. If increased competition in wireless access can help decrease costs, the increased service penetration that follows should lead to more opportunities for economic and political participation.

Social and Technological Modernization

Technology changes culture. That much is clear from the experience most of us have had with the introduction of new technologies in our lifetime, such as the telephone, television, and the computer. There exists a whole wealth of literature on the relationship between technology and culture, ranging from the examination of printing presses and stirrups, to the development of the city, the sciences, and the technologies of the modern age. There is less consensus on what kinds of consistent patterns can be drawn from these diverse examples. Nevertheless, the political and economic scientists have largely agreed on a term for the process: modernization.

Generally speaking, social, economic, and technological modernization are thought to be connected, but it is difficult to determine the causal linkages. Joseph A. Schumpter, for example, divides his "modern" and "nonmodern" classes by technological use more than simple income levels because of the large pockets of high technology that drive production and politics in many parts of the developing world. Without attempting to assign causal links, the conversation usually begins with technology for a number of reasons. From the perspective of theory, the most common formulation is that technology drives economic growth and productivity; but, as the developmental theorists in economics and political science correctly point out, growth and productivity do not necessarily equal development.

This theoretical clarification is especially critical as the world economy continues to move toward a larger service sector and the importance of other sectors diminishes in proportion to the service economy (Bell, 1973). The economic revolution underway today will be driven "not by changes in production, but by changes in coordination" (Malone and Rockhart, 1991). What we are witnessing is a new kind of control revolution brought about by distributed information systems and global communications networks (Neuman, 1991).

Many economists would use the term coined by Simon Kuznets to describe these technological developments: We are in the midst of an epochal innovation, which, to his way of thinking, is the key to modern economic growth (Kuznets, 1953). Here we can parrot some of the wide-eyed descriptions of the popular media, which range from the "information age" and "information superhighway" phraseology, all the way to the "cyberspace" jargon of popular culture.

The political scientists, too, have focused a great deal of intellectual energy on describing the institutional framework of modernization. Most commonly, they have described the modernization element of the development process in terms of stages. Perhaps the most famous metaphor that surrounds talk of development is Walter Rostow's concept of the "take off" point and A. F. K. Organski's stages of development. By assembling the requisite resources and bringing the appropriate institutions to bear, a country can "take off" and "drive to modernization."

All of these classifications are politically charged. *Modernization* and *development* are not neutral words; they are connected to certain kinds of social structures and economic practices prevalent in the developed world (Porat, 1977). It is not considered a modern practice to bar women from holding political office, for example. In many countries in the Middle East, though, the high technology of oil production coexists side by side with the traditional practices of male dominated politics (Hess, 1994).

Summarizing both the staged descriptions of the political scientists and the technological emphasis of the economists, Samuel Huntington (1968) cast the issue of modernization in a broadly social frame:

> Those aspects of modernization most relevant to politics can be broadly grouped into two categories. First, social mobilization means a change in the attitudes, values, and expectations of people from those associated with the traditional world to those common to the modern world. It is a consequence of literacy, education, increased communications, mass media exposure, and urbanization. Secondly, economic development refers to the growth in the total economic activity and output of society. It may be measured by per capita growth, national product, level of industrialization, and level of individual welfare gauged by such indices as life expectancy, caloric intake, supply of hospitals and doctors. Social mobilization involves changes in the aspirations of individual, groups and societies; economic development involves changes in their capabilities. Modernization requires both. (Huntington, 1968, p. 35)

Part of the problem in understanding these two transformations sketched out by Huntington is that commentators often focus on modernization as an undifferentiated force for good that penetrates a society along the pattern of a traditional diffusion curve. But the simple fact of the matter is that all good things do not go together, and modernization does not have to start, and seldom does begin, from a single point. What research from the field has made clear is that certain segments or classes of society begin by adopting a certain social practice or technology, they are seen as leaders, the leaders are eventually followed by the rest of a community, and the community is thereby transformed (Rodgers, 1962).

But who is to lead? There tend to be two leadership models for "modernization." One is state sponsored and most often associated with the left wing of the political spectrum. To a great degree, the fall of the Soviet Union has discredited many of these kinds of development schemes, although the economic success of many countries in Asia do, in great part, depend on the active involvement of the state in economic and social planning.

The second model is decentralized, starting with local communities and defined by the needs of those community. Some of these decentralized visions of modernization come from the American experience of grass roots political and economic development, but other strains of this thought resonate in corners of the political world from Chipas in Mexico to the regional growth patterns in Russia.

Modernization through investment in wireless telecommunications infrastructures has predominantly, if not exclusively, been based on the first model (Davies, 1994; Mueller, 1993). But there are other visions of modernization which may be appropriate to the telecommunications sector. Some of the most compelling visions of modernization driven by technological diffusion comes from the established literature which connects social policy with appropriate technology (Schumacher, 1973). There is also the issue of forward and backward linkages for technological modernization to be considered; how can a decentralized model for sustainable telecommunications investment be articulated so that appropriate connections are made to customers both in the service territory and throughout the world? (Hirshman, 1970).

This literature suggests an opportunity to balance state involvement with a distributed, decentralized model for national development. Modernization can be driven through appropriate technology that fuels energy at multiple locations throughout a community, allowing the goals of economic and political development to be achieved through a decentralized pattern of community empowerment. If such a design can be connected to a liberalization policy, we could begin to construct a coherent map for achieving the social, economic, and political goals associated with modernization.

Because of its scalability and lower fixed costs, wireless infrastructures are potentially a much more powerful tool for decentralized models of telecommunications development. Decentralized development requires a degree of flexibility, allowing for the quick expansion, retrenchment and reallocation of resources that a wireless architecture can give. Appropriate strategies for backbone integration, be it point-to-point microwave or wireline connections from each mobile switching office, can offer the forward and backward linkages to maximize the social and economic viability of the network. And, from the perspective of a corporate manager looking to serve the needs of poorer populations in developing countries, it is a heck of a lot easier to set up a cellular or fixed wireless local loop system in the barrios of Bogota or the slums of Manila than it is to lay a wireline one.

Effective and Appropriate Public Policy

One of the assumptions behind development thinking is that there is a role for public policy and government involvement: There are things we can do to make the development process work more smoothly.

But the present rhetoric on the role of public policy worldwide focuses almost exclusively on the less or more aspects of government, rather than on issues of effectiveness. The issue of what to make policy about gets lost in the shuffle, especially as it pertains to the telecommunications sector.

The groups on the right side of the political spectrum generally are the ones arguing for reducing the role or abolishing outright existing telecommunications regulatory bodies. The techno-libertarians tend to be the most strident, often arguing that institutions like the FCC can be abolished in favor of the competitive checks and balances of the marketplace. Those in the strong liberalization camp concentrate on using existing corporate law and regulatory institutions that apply to all sectors of the economy, minimizing the differences between the operation of the telecommunications sector. On the left, it is largely assumed that regulatory bodies will live on in perpetuity, and that the political structure should take an active role in protecting and developing the telecommunications sector. Overlapping protections, regulation on both the content and nature of information carriage, are staples of the policy proscriptions offered by the "protectionist" and "subsidization" camps.

For public policy to be successful in the development process, it is commonly assumed that government must bring resources to bear that would not be available otherwise. The opportunity for new kinds of participation and a decentralized form of modernization can alter how public policy informs the liberalization process: By emphasizing the sustainability of competitive providers, the institutions of government can help organize resources that presently exist effectively instead of struggling with the apparent lack of resources.

One of the critical assumptions of development thinking has been that the state is the central focal point for social action (Skocpol, 1979). Economic and political development theorists share one conviction: Planning is essential. We can add to that another truism of public policy: Success is survival. The burden of development rests on the institutions of public policy that are thought of as the directing force for development. Through revolution or election, leaders who do not provide for development are removed; even in the case of the communist bloc, for so long seen as unshakable, the lack of economic prosperity and development eventually brought down the governments (Brzezinski, 1990).

But planning is not easy, especially in developing countries that are often characterized by the "weak states" that govern them. Developing countries often do not have the intellectual and physical resources necessary to perform the long-term strategic planning that many developed countries have integrated into their governing structure.

At the center of these successes and failures are the institutions of management: the government and the corporation. When it comes to the strategic management of infrastructure resources, the responsibility has traditionally fallen to the public sector. The main reasons for this are both theoretical and practical. Economists have identified economies of scale that would be appropriate to monopoly production and have advocated, in some cases, the strong regulation of a monopoly provider for infrastructure services. At a practical level, much of the capital required for the construction of such projects would have to come from deep pockets, and, no matter how inefficient they may be, there are few deeper pockets in world history than modern Western governments.

The origins of the electric, gas, and telecommunications industry were all dependent on direct government intervention, and governments have traditionally used their role in such projects to maintain control over resources that were thought essential to the well-being of the nation. During the cold war, one of the best rehearsed arguments in favor of retaining the monopoly structure of the Bell System was that AT&T represented the most critical asset in the maintenance of national security.

Even in developed countries, the existing monopoly cultures make change difficult, if not impossible. Established interests will continue to collude and cross-subsidize services in order to retain a monopoly position. But in many cases, the policy institutions are not independent from the providers of services and are not willing to take a proactive stance in the introduction of competition (Tyler and Bednarczyk, 1993).

Public policy will still have to be focused on managing social assets and resources, but the kind of management that will be required is changing rapidly. The traditional explanations of how and why policy is important to the management of infrastructure sectors no longer is sufficient guidance for corporate and public policy makers. A new understanding of how the public and private sectors can cooperate will be essential to the construction of new institutions that match competitive realities.

How communities adapt and adopt new institutional frameworks will determine, in great part, how successful they are in providing these essential services to their community. As the World Bank notes, "the performance of infrastructure [services] are derived, not from general conditions of economic growth and development, *but from the institutional environment*" (World Bank, 1994).

For the telecommunications sector, the two pieces of work that relate most closely to this issue come from Raymond Duch (1991) and Ben Petrazzini (1995). Duch's argument is that, for the three developed world countries he examined, institutional environments which favor political pluralism and manage interest groups well are the ones most likely to successfully privatize telecommunications networks. Petrazzini's work focuses on the developing world, and his argument basically sketches out what appears to be the opposite reality in poorer nations: The possibilities of reform and privatization depends on a strong political authority in the center. These two works are touchstones for the remainder of the discussion and will come in at various points in the case study chapters. But what is clear from just the summary of both works is that connecting the agenda of liberalization with the institutional environment is critical—especially for any public or private sector manager who believes the future of telecommunications depends on the implementation of liberalization and privatization policies.

This brings us back to the concept of a "weak state." A government without the institutional means to effectively allocate resources is left adrift, with few means of guiding social or economic development. Quite often, that requires the political leadership to use the military as a crutch to support themselves in power. Yet there are some instances where countries with limited resources and no institutional framework to guide development have succeeded.

It is a quality of political and economic leadership that has helped many developing countries move forward; observers point to the successes of Korea, Singapore, Taiwan, and the other newly industrialized countries of Asia with a view to promote their strong sense of leadership. Whether or not the "Asian values" of Lee Kwan Yew are to be emulated in other countries is not the issue for our purposes. It suffices to say that a strong leader or leadership group that can help manage what are limited resources becomes an essential asset in the struggle for economic and political development.

This is also true in developed countries, but in a very different sense. There are plenty of experts willing to testify in front of a congressional committee that a particular policy is a good idea. There are hundreds of consulting organizations that are willing to explain to a corporate manager why he or she should follow a specific policy. But many countries lack the leadership required to make informed policy decisions because they are unaware of the resources that actually exist (Thompson, 1993).

Liberalization needs to reconstruct the role of public policy as the mechanism to identify resources that might help sustain competitive institutions. In incorporating this value into an overarching policy framework, strategic liberalization argues that there is room in the center for a more proactive role for government institutions and regulatory bodies worldwide. Instead of concentrating on restricting and constraining activities, a variety of policies can be implemented to further liberalize and develop portions of the sector that are of strategic importance. Because wireless communications shows promise as a driving force for increased technological penetration and social modernization, it will be critical to define an appropriate role for public policy in this specific area.

But what kind of policy would be appropriate? And how can that policy incorporate the needs and interests of corporate and government institutions? These questions are the focus of the remainder of this book as the parameters of how to use wireless communications as an engine of growth, competition, development, and liberalization are outlined.

CONCLUSIONS: STRATEGIC LIBERALIZATION AS THEORY

Strategic liberalization, as a policy for infrastructure development, takes advantage of the social, economic, and technological changes that have driven the alterations in the provision of infrastructure services worldwide. A definition has already been provided: The implementation of specific policy measures to increase competition in the market for wireless access services, such as cellular, wireless local loop, and satellite communications. A strategic focus on wireless communications will provide a sustainable foundation for the continued growth of the telecommunications sector and the successful introduction of new products and services in an environment of facilities-based competition. In addition, it is a policy that can help public and private sector managers identify economic and technological trends and adapt political and economic institutions to facilitate development through participation, modernization, and the reinvention of public policy.

The word strategic is meant to emphasize the fact that corporate and public managers have strategic choices to make in order to ensure the implementation of a successful policy. Strategic choices about particular technologies, for example, are critical to the success of liberalization. In the telecommunications sector, as we discuss, there are a range of possible technologies and services that can be emphasized in the first stages of liberalization. Traditionally, liberalization has focused on the high-cost technologies, such as value-added networks, leaving the public network monopolized. That choice can now be inverted; liberalization can start with low cost wireless technology that would have an immediate impact on the establishment of competitive markets.

Strategic choices have to be made about the structure of the governing institutions. Should regulatory regimes concentrate on constraining large, sometimes, monopolistic providers, or should they take proactive steps to diversify the marketplace? In most cases, regulatory agencies have acted as watchdogs, reacting to marketplace changes instead of anticipating them. A more proactive stance is possible and is certainly needed during this period of change. But which areas and technologies are appropriate choices for investment and liberalization?

Corporate managers have strategic choices to make. Traditionally, corporate managers have used established services as a base for determining the cost and structure of future services, but for the provision of infrastructure, such an assessment is difficult, if not impossible. There are no established services that would allow us to conduct a traditional customer-needs focused marketing study to define products and services. So the questions are evident, even though the answers are less so: What is the size and scale appropriate to the provision of certain services? And what kinds of service architectures will allow a company to maintain a sustainable investment policy in an environment of facilities-based competition?

During the course of this chapter, the basic themes and issues of the book have been addressed in general terms, outlining the idea of strategic liberalization and its grounding as a theory for infrastructure development. Now it time to turn to more a specific and detailed examination of how that theory would be implemented in practice. In great part, this means showing how the values of development we have outlined could find specific expression in particular institutions and certain kinds of government and corporate activity.

But, with the agenda of privatization reaching completion throughout the world's telecommunications sector, corporate and government managers unsure of their footing. Have the goals of privatization been achieved? Yes and no. Yes, inasmuch as the ownership of public sector companies has been transferred to private hands. The transfer has often met the two traditionally defined goals of development mentioned before: It has increased participation by the distribution of wealth and by allowing for further private sector contributions, and it has provided the capital needed for continued modernization of services.

This returns us full circle to the issues that appeared at the beginning of the chapter. Privatization is an unfinished business, and liberalization is the best opportunity to fulfil the promise of privatization: lower costs, better service, and national development. But a policy of deregulation and privatization is not in itself a policy which will improve the provision of services. A different kind of market liberalization is required if privatization is truly to achieve the aim for which it was intended: improved services and truly sustainable development of infrastructure resources.[9]

Providing a framework to answer these questions requires a assessment of technologies, combined with an examination of the social consequences of these technologies as they relate to their potential as profitable products and services. The grounding in the social sciences which strategic liberalization offers provides a further dimension for analysis that is not included in the traditional perspectives of telecommunications policymakers in the public and private sector.

In addition, the policy of strategic liberalization defined in this book attempts to bring a common character to the choices that are being made by corporate and public managers. Both corporate and public managers hope to establish institutions that are microeconomically and micropolitically sustainable and that have the ability to compete for the provision of services. By describing the opportunities for decentralized modernization and enhanced participation, and by pointing to specific occurrences in the four countries that we will be examining, our goal is to help define as a global community how the quality of and access to infrastructure services can be improved.

The next two chapters further outline the dimensions of strategic liberalization and what it might mean for the telecommunications sector. Specifically, the chapters examine the possibilities for wireless communications to provide the foundation for further competition in the telecommunications marketplace, thereby creating an opportunity for public and corporate policy makers to use wireless access as a cornerstone for future competitive planning. Chapter 2 examines the development of the telecommunications sector in historical perspective, analyzing the different types of competitive pressures that have evolved and their potential impact on development strategies. Chapter 3 focuses on wireless communications, discussing the altered cost structure and policy environment of wireless access, reviewing the character of existing competition, and discussing the strategic potential of wireless access as a driving force for telecommunications development.

In order to ground this discussion in the real world of telecommunications development, chapters 4 through 7 speak to specific conditions in four case study countries, namely United States, the United Kingdom, Russia, and Brazil, and how the policy of strategic liberalization has been developed in each. The final chapter draws conclusions from each of the case studies and offers a comparative discussion about the possibilities for future policy.

The old ways of doing development are no longer viable in the changed political and economic environment. Political and economic participation is not what it once was. Modernization now means something very different. Unless we breathe new life into these terms and thoughts, they will become irrelevant, leaving people who might benefit from such ideas marginalized. Strategic liberalization represents a new and critical opportunity to regain the initiative as we continue to search for a better future for the citizens of the world.

ENDNOTES

[1] The jungle analogy is borrowed from the conclusion of George Calhoun's *Wireless Access and the Local Telephone Network* (Norwood, MA: Artech House, 1992), and is informed by many recent efforts to more directly link theories of economic and technological development to models of biological and ecological change. Some of the most notable include Michael Rothschild's *Bionomics: Economy as Ecosystem* (New York, NY: Henry Holt and Company, 1990) and Joel Mokyr's *The Lever of Riches: Technological Creativity and Economic Progress* (Oxford: Oxford University Press, 1990).

[2] Generally speaking, privatization is justified if the "sum of the social value of the enterprise in private hands plus the social benefit of the monetary transfer from the private to the public sector associated with the sale proves to outweigh the social value of the enterprise in government hands." From *Privatization in Latin America*, Manuel Sanchez and Rossana Corona, Eds., (Washington, DC: The Johns Hopkins University Press, 1993), p. 2. See in particular L. P. Jones, P. Tandon and I. Vogelsand, *Selling Public Enterprises*. (Cambridge, MA: The MIT Press, 1990) The "inconveniences" caused by political management of economic institutions are also a compelling justification for privatization policies.

[3] Raymond Duch, *Privatizing the Economy*, p. 2. Duch writes:

> Looking to political variables to explain variations in the performance of government owned firms only makes sense where an economic explanation is clearly insufficient. There would be little unexplained variance if most government-owned firms performed poorly and most private entities performed well. This is not the case; a considerable amount of variation in the performance of firms cannot be accounted for by the ownership variable. How, for example, do we explain the dismal performance of such large French companies as Thompson and Rhone-Poulenc under private ownership and their dramatic turn-

around under nationalization? Why do French banks continue to perform well under nationalization? How do we account for the significant improvement in the performance of the British nationalized sector in the late 1970s and early 1980s—British Steel being one of the prime examples? The point here, of course, is that we need to look beyond economic explanations that focus simply on ownership.

[4] In the next chapter, we examine the particular arguments for the monopolistic provision of telecommunications services throughout the world, but for this stage in the discussion it is the consequences of the general argument that are most important. From a microeconomic point of view, the structure of a monopoly firm is based on setting prices that reflect the value of the product demanded and producing accordingly. The critical issue that we are ignoring at this juncture is: Where do monopolistic companies get the information required for pricing certain products? Because there is no market mechanism, information has to be from other sources. Most significantly, the government has played the role of "price setter" in many of the monopolized industries throughout the world. See, in particular, William Baumol et al., *Contestable Markets and the Theory of Industry Structure* (New York: Harcourt, Brace and Jovanovich, 1982). The institution created in a monopolistic arrangement is therefore based on the assumption that a centralized source of production and management is best.

[5] The problem of identifying collusive behavior versus competitive behaviour in the private sector remains one of the most studied and contentious issues in modern economic theory. For an examination of some of the fundamental issues in this field, see, in particular, George Stigler, *The Organization of Industry* (Chicago: The University of Chicago Press, 1968). One possible way to address the difference between real competition and collusive behaviour is to determine the variance in economic strategies and results of strategy implementation between firms within the same sector. See Omar Toulan, "Sources of Local Variation and their Implications for Strategy," Unpublished Manuscript at the Alfred P. Sloan School of Management, Massachusetts Institute of Technology (1994)

[6] For example, because credit institutions are not very deep, or are nonexistent in many parts of the developing world, what are some of the social regularities that would be required as the foundation of credit institutions and markets? That is the question at the heart of Robert Putnam's recent work on "social capital," which he defines as the ability of groups of people to construct the open environment required for strong institutions. A variety of social commentators have recently picked up the mantra of "values" as a key to social and economic development, including Ben Wattenberg, Francis Fukuyama, and William Bennett.

[7] Harold Innis, *Empire and Communications* (Toronto: University of Toronto Press, 1951). Innis argues that the structure of politics and economics in any society is based on the structures and mediums of communication. For example, communication by paper and pen has certain structural characteristics that differentiate it from the printing press; Innis claimed that the paper and pen of the monastic culture had a direct impact on the closed and centralized political and economic structures of Europe during the middle ages. Shifts in communication from one medium to another also shift political structures. The advent of

the printing press altered the nature of communication, which in turn undermined the monastic, Latin-based culture and replaced it with nation-state organizations. For further information on this particular transition, see Elizabeth Eisenstien, *The Printing Revolution in Early Modern Europe* (Cambridge: Cambridge University Press, 1991).

[8]   Al Gore, *Scientific American*, March 1991. This issue is especially critical considering the recent debate in the United States on the relationship between genetics and cognitive tests, as it appears in Richard J. Herrnstein and Charles Murray, *The Bell Curve: Intelligence and Class Structure in American Life* (New York: The Free Press, 1994). There are clear reactions in the developing world to the widening gap in economic and educational opportunities across class and racial divides. For an effective and, to my mind, devastating critique of the arguments in the book, see Leon J. Kamin, "Behind the Curve," from *Scientific American*, February 1995.

[9]   W. Russell Neuman and Lee McKnight are responsible for much of the ongoing work on the possibilities for Strategic Liberalization. The phrase was coined by Neuman as part of an extensive analysis of the Telebrás ownership and provision structure in Brazil. The work on that and other projects has been the stimulus for much of the telecommunications theory articulated in this book.

The Transformation of
Telecommunications

The "jungle" of the telecommunications sector is quickly transforming itself. The evolutionary process can be mapped out in terms of years, rather than decades or centuries. Keeping abreast of the changes in the market requires a global, as opposed to national, focus: Markets and companies are beginning to intertwine and intermix, a fact that will produce even more varied "species" of telecommunications providers throughout the world.

As the sector evolves, the framework that links telecommunications development to economic growth and improved prospects for political and social participation will also be altered. The classic lines that have been drawn to map that specific relationship will become increasingly less relevant, requiring managers in both the public and private sector to look deeper for new opportunities to link services and capabilities with people.

This chapter examines the transformation of the telecommunications sector from a global perspective, and discusses the implications for development. The first step is to describe the early progress of telecommunications services throughout the world, discussing some of the common technological, economic, and political factors that have shaped the provision of services. The next is to examine the impact of these institutions on the development process, examining some of the causal models that connect increased telecommunications investment to national development. Finally, the discussion leads to the new technologies and kinds of competition that are most likely to contribute to the goals of national development. By the end of the chapter, we can turn our attention on the role of wireless communications services as a critical opportunity for implementing a policy of strategic liberalization.

DEFINING THE CHARACTER OF
TELECOMMUNICATIONS SYSTEMS:
A HISTORICAL PERSPECTIVE

The companies and government ministries that provide telecommunications services today offer customers the opportunity to communicate through certain kinds of technology. The technologies are diverse: Telephones, televisions, satellites, computers, electronic mail networks, personal digital assistants, just to name a few, and more are being introduced every day. The content and nature of communication is limited only by human thought and ingenuity. From basic conversation to the collaborators on the Internet, all different kinds of content will be circulating on the networks of the future.

But much of that is hype. Fortunately, we can ask Ithiel de Sola Pool (1990) to describe, in simple language, what is going on when we speak of a communications system:

> Viewed physically, a communications system consists of 1) a series of nodes, or terminals, each of which is an input device, or an output device, or both; 2) a transmission medium among the nodes; 3) sometimes a switching device that determines which nodes are connected to which; and 4) sometimes a storage device for holding messages and forwarding them later on. It can be a one-way communication system—like broadcasting—in which one node talks and the rest listen; or it can be a two-way communication system like the telephone. (Pool, 1990, p. 19)

At the fundamental level, telecommunications is the art of connecting people through a medium. How that happens is the critical issue in discussing the differences between kinds of communications.

The traditional telecommunications company has played the role of "transmitter," carrying signals from one place to another. Two means for transmitting communications have been developed: through wires, which allow for the transmission and reception of information at the fixed points along the wire, and through wireless, which uses the electromagnetic spectrum to carry information from one place to another.

In that regard, we can characterize the service provided by these institutions as *access to communications.* By stringing a copper wire to a person's telephone, or handing a person a cellular phone, or by connecting a person's television or a satellite or cable television access wire, the company offers a person access to a certain kind of communications. How access is provided, the quality and cost of access, and the nature of access are all bound up in the economic and social developments of the last 200 years. This brief history of telecommunications explains some of the dynamics that have shaped the kinds of telecommunications services that are available throughout the world today.

Early Developments in the Telecommunications Industry

Telecommunications does not start with Alexander Graham Bell. As a practical man of science, Bell's work depended on a range of scientific innovations in physics, chemistry, and other physical sciences through the course of the centuries preceding his invention of the telephone. The development of the telegraph was closely tied to innovations and experiments in the transmission and production of electricity, for example.

But to understand the place of the telephone and other telecommunications technologies, we need to go back further than Bell's invention. For centuries, governing authorities had established systems of communication to link the often far-flung territories they controlled (Anderson, 1983). Most of the efficient systems relied on the visual transmission of signals from one station to another (Derry and Williams, 1960). With the expansion of electrical and transportation systems during the wake of the industrial revolution, a new control structure emerged for the coordination of these activities (Beniger, 1986). At the core of this information infrastructure was the manipulation of electric current in the form of the telegraph.

The first experiments in telegraphy were undertaken in the mid-18th century, but the problems of controlling and harnessing electricity were not resolved until the experiments of Michael Faraday in the early 19th century. A variety of devices were created to code and send messages through electric pulses, transmitted in one place and received in a second (Frederick, 1993).

It was left to an American to make the experimental breakthroughs a commercial success. By taking a simplified version of the two-wire telegraph, standardizing the dot-and-dash codes, Samuel Morse established a commonality that could bring far-flung points together. On January 1, 1845, Morse's machine successfully sent the message "what God hath wrought" from Baltimore to Washington. A year after his success, 1,445 kilometers of telegraph lines were active in the U.S.

By 1862, the world's telegraph system covered approximately 150,000 miles, and telegraphy was established as a critical technology of coordination for the industrial age. Soon after that, Cyrus W. Field succeeded in linking the Americas and the European continent with undersea cables, allowing for a message transmission from President James Buchanan in the United States to Queen Victoria of England (Jerpersen and Fitz-Randolph, 1981).

The basic technology of telegraphy was transformed into a control structure in a time of increasing industrialization throughout the developed world. Business and commerce began to rely on the possibility of two-way electronic communication to direct and guide strategic decisions and coordination. Automatic switching and signals for the railway system were installed and completed from the 1860s to the 1880s, completing the "leveling of times and places" that such technology made possible (Beniger, 1986).

The telegraph was so well established that, when Alexander Graham Bell applied for a patent on his working model of the telephone on February 14, 1876, few would have predicted it would first challenge and then replace telegraphy as the dominant form of two-way communication throughout the world—especially considering the fact that the idea of transmitting voice through a wire by means of producing variable resistance in the microphone and consistent resistance through the amplitude modulation of an electric current in the wire was not new. Concurrent with Bell's invention and commercial licensing of his patents, similar devices were registered in Britain, France, and Germany.

Stretching a wire from one place to another and using that wire to transmit information allowed for a limited sense of communication. There were two options: Either the wire started in one place and stopped in another, in which case only one person could communicate with no more than one other person, or one wire could connect to many people, in which case only one person could "communicate" and everyone else would have to wait to communicate with everyone else. For that reason, the initial applications of the telephone were for what would be called today, "broadcasting" (Pool, 1977). Much like cable television today (though certainly not like the cable television of the future), a central source that sounds through the wired network to all the receivers who subscribed to the service. In fact, daily weather reports and concert symphonies were the order of the day in many continental telephone systems before the turn of the century, complete with daily schedules and regular broadcast formats.

The critical invention required to transform the telecommunications industry at this stage of development was the switch. Before switching, the telephone network did not have the ability to route messages and conversations from place to place so as to ensure that each person with a telephone could call any other person with a telephone. With switching, connections could be made between points and two-way conversations could ensue. The first switches were human operated, and the development of an electromechanical switch did not occur until early into the next century. Even then, the telephone systems of the developed world relied on human operators and switching until the 1930s and 1940s.

Following the expiration of Bell's patent in the United States, a period of competition ensued that dramatically increased the level of penetration of telephony (Mueller, 1993). The Bell System, guided by Theodore Vail and backed by the financing of J.P. Morgan, did battle with a number of local telephone providers. Meanwhile, the European powers had decided that telegraphy and telephony posed a direct threat to their monopoly of all postage services. As such, many of the countries began to bring all telephony services under the rubric of their Ministries of Posts, thus creating the Post, Telegraph, and Telephone administrations that have dominated the provision of such services in Europe for the course of the 20th century (Noam, 1992).

Throughout what is commonly known today as the developing world, reasonable levels of telegraph and telephone penetration were achieved. Telegraph and telephone service was inaugurated in many Latin American countries soon after its introduction in the United States and Europe. Colonial administrations extended the new communications technology in the continuing expanse of imperialistic power towards the end of the 18th century.

During this expansion of wired telecommunications, Guilermo Marconi was experimenting with a completely different way to send messages and voice from one point to another. Following the experiments of the physicists that set the foundation for Einstein's theory of relativity, Marconi developed a system for "wireless telegraphy." The early development of wireless was marked by the politics of the age; the first applications that became profitable for the Marconi company were the ship-to-shore services that the British, Italian, and German navies prized so heavily during the years leading up to World War I. Marconi's services were not allowed to directly compete in Great Britain with the telegraph and telephone services administered by the British Ministry of Communications. As such, many of the applications that have become so significant today, such as mobile telephony, were delayed for another time and another age.

The Age of the Public Utility

With the final reforms of the communications sector in the early 1930s in the United States, a standard organization for the provision of telecommunications services had been established worldwide. Centralized organizations, grounded in government ministries or what became known as "public utilities," defined the character of telephone service. That fact is clear in the history books. Why it happened is still a source of significant debate.

The traditional view is best expressed in the work of Alfred Chandler (1990), an economist and organizational theorist that many draw on for his particular analysis of the telecommunications industry in the United States. His view relied on a definition of economies of scale that is used to justify the existence of public utilities; when the average cost of producing another unit of output falls with each unit of output, an external economy of scale exists. Other producers in the marketplace would cause economic waste inasmuch as a single company could produce the output at a lower average cost than the average cost of two or more companies.

But Chandler's argument relied on more than just a traditional description of scale economies and their effect on production. He also argued that a certain kind of organization was required to support the provision of telecommunications services worldwide. Chandler's view is that:

> The speed and volume of messages made possible by the new electric technology forced the building of a carefully defined administrative organization, operated by salaried managers, to coordinate their flow and to maintain and expand transmitting facilities. The first enterprise to create a national organization to handle through traffic obtained an almost unassailable position. (Chandler, 1990, p. 202)

The combination of organization and technology produced a natural economy of scale. Monopoly organization for the telecommunications sector was natural and therefore enshrined in the public policy of the age.

Others see it differently. Milton Mueller (1993) argued that competition was the real stimulus of telephone penetration in the United States. Through a process he called "access competition," telephone service was extended to rural and small-town America by small- and medium-sized companies hungry for a share of the market for local telephony. The small companies were able to develop and maintain the kind of system they did because the economies of scale in the telecommunications industry come more from switching than from actual transmission. "The small-scale telephone switchboards needed by small towns and rural areas were easy to manufacture and inexpensive to operate," wrote Mueller, thereby creating the conditions for competition and increased penetration.

Mueller's argument brings into question the view that increasing penetration of telephone services nationally was only possible through a monopolized, in many cases publicly owned, telephone company. Andrew Davies, employing a methodology founded in political economy, sees the influence of American financiers and centralized politics in Europe as being more the determining factor than the kinds of technological and organizational dynamics described by Chandler (Davies, 1994).

The Structure of the Public Utility

This is not the place to rehearse those debates and bring them to a conclusion. But it is fair to say that the creation of the public utilities of this century was a political decision. There were other options for the development of infrastructure services such as telecommunications, but these paths were not taken. Regulatory agencies and ministries were constructed to closely monitor the establishment of service, along with the cost and quality of that service.

The nature and structure of public utilities settled in the first quarter of the 20th century. In the United States, that meant a compact between a private company, American Telephone and Telegraph, and the government. Embodied in the Kingsbury Commitment of 1913 and the subsequent Communications Act of 1934, AT&T was given the nearly exclusive right to provide telephone service throughout the country. The Bell System provided that service through a group of Bell Operating Companies, which integrated their service with the long lines department and the equipment producing division of AT&T, Western Electric. As part of that public compact, though, the Bell System had to offer service in conjunction with a number of smaller companies that had developed during the period of competition from the late 1880s to the early part of the next century.

Overseeing the provision of services in the United States was the Federal Communications Commission, established as part of the 1934 Communications Act. State utility commissions, which had become an important part of state regulation of utility activities by the turn of the century, took on the responsibility of overseeing the provision of services in their respective jurisdictions (Teske, 1992).

In Europe, there was no need for such a compact inasmuch as the government had taken early control of the provision of services. Additionally, the complications of a federal system did not exist in the unitary states of Europe at the turn of the century. The establishment of Post, Telephone, and Telegraph (PTT) administrations created the European version of the public utility. The government ministries were mostly self-regulating, with the management of the telecommunications systems being the responsibility of politically appointed managers.

This structure was, in great part, a kind of entente between the government and a variety of private investors wishing to offer infrastructure services. The entente fixed the structure of these institutions for more than 60 years, with very little structural change. The stability provided an opportunity to increase penetration of services, and the regulations put in place for defining the activity of these institutions reflected this opportunity.

Rules and Regulations Guiding Utility Institutions

In this stage of telecommunications development, service offerings of for all of the companies around the world looked pretty much the same: The public, switched telephone network brought you telephone service. The network was "public" because the goal of the service was that everyone should be able to gain access to the service, and "switched," inasmuch as it was an integrated network that linked all subscribers together through the technology of the switch. Even though the numbers were different and the kinds of dial tones or rings were also different, telephones, from a functional point of view, were telephones. You spoke through on one end, and on the other end, the person heard your voice.

The guiding principle for the public utility was the common law concept of common carriage, imported to the United States and much of the Western world from the British legal tradition (Noam, 1994). The concept of common carriage evolved with regard to what we would today call infrastructure services, such as transportation and communications. Because of the nature of the service and its importance to commercial and business transactions, a common carrier owes a duty of nondiscrimination; the "common carrier was required to serve, upon reasonable demand, any and all who sought out their services."

The common carrier system has largely defined the character of regulation as it relates to the provision of telecommunications services in the developed world. As Eli Noam pointed out:

> The common carrier system has served telecommunications participants well: it has permitted society to entrust its vital highways of information to for-profit companies, without the spectre of unreasonable discrimination and censorship by government or private monopolies; it was an important element in establishing a free flow of information, neutral as to content; it reduced administrative costs and the burden of liability of a carrier, since it needed not, at least in theory, inquire as to a user's background and intended use; and it protected the telephone industry from various pressure

groups who would prevent it from offering service to their targets of protest or competition. (Noam, 1994, p. 437)

Under such protection, the telecommunications industry was able to flourish
in countries with sufficient capital to develop extensive telecommunications networks. The combination of standardized, undifferentiated telephone service and a
common carrier network became the defining points for the concept of "universal
service." First defined more as a public relations term by AT&T at the turn of the
century, universal service quickly become a fixture of American policy. In the
1934 Communications Act, the goal of universally accessibility was set into law
and remains a firm goal of communications law to this day.

The problem was clear: What kinds of institutional activity could best support
that goal? In the United States, a number of subsidized companies sprang up to fill
in the gaps where the Bell System companies could not economically recoup their
investment.[1] Where the companies were publicly held, more often than not a direct subsidy was given to those individuals who were unable to purchase telephone service or who lived in geographically isolated regions.

In the developed world, universal service has largely been realized through the
public utility system and PTT administrations. Penetration rates of 93% to 95% in
some countries indicate that the telephone has become a pervasive instrument of
our society, connecting millions of people each day through a service that has become regular, reliable, and thought of as essential.

Perhaps most significantly, universal service was achieved largely at the expense of some of the large, corporate institutions that had an intensive need for
communications services. Local telephony was basically subsidized by long-distance service, which was priced well above cost in almost all of the developing
countries. The reasons were mostly political. Lower rates for local consumers and
"essential services," such as information, police, and fire services, were very popular with citizens and thus a boon to elected officials.

But as penetration rates increased in the developed world under the aegis of
public utility and PTT management, the telephone did not fare as well in poorer
countries through the world. Many of the states that formed during the period of
decolonalization choose the dominant style for the administration of their telephone administrations; government ministries cropped up throughout Africa and
the former Asian colonies as telecommunications service providers. Telecommunications providers in Latin America, many of which were independent even into
the 1960s and 1970s, were nationalized for the purposes of achieving higher rates
of performance. Nevertheless, in the developing world, telephone penetrations
most often are below the level of 10 per 100 people. Considering the lack of resources and the traditional difficulties that these telephone administrations have
faced, universal service seems to be an impossible task.

The Tragedy of the Public Network:
The Breakdown of the Public Utility Model

Whether or not a natural monopoly ever did exist, or whether it was mostly a theoretical construction used to justify a political decision, the public network did thrive from its establishment at the beginning of the century into the 1960s and 1970s in the developed world. Then, decreases in prices and new technological opportunities began to change the public network.

The first changes were seen among the intensive users of telecommunications services, most particularly large corporations. In the words of Marcel Roulet, then chairman of France Telecom:

> We in the business community have seen a new type of corporation emerging. We call this the global enterprise, and it is marked not merely by its capability to sell goods or services on an international scale, but its ability to coordinate the activities of international operating entities into a smoothly-functioning unit, regardless of distance or differences in time, language, monetary structures or culture. (Coopers & Lybrand, 1993, p. 12)

Of course, the global enterprise may only be built, operated, and maintained through the use of a global enterprise network. This is no secret, and in fact the corporate strategies of many of the world's largest corporations have for the past few years been focused on creating the most effective enterprise networks, that will allow them to achieve their strategic aims, anywhere on the globe.

The impact on global communications has been revolutionary. For corporations, making a call halfway around the world needed to be almost as easy as making a call down the street. Global communications networks catalyzed this development, facilitating the transformation of the telecommunications providers and the customers they served through the deployment of new infrastructure capacity.

The results were apparent in the statistics. Usage skyrocketed. Calling minutes, for both long distance and international calls, have increased dramatically since the early 1970s. Not only have businesses increased their usage of telecommunications services, so have common citizens who have benefited from lower prices and the opportunity to reach out to this new global community (Staple and Mullins, 1991).

But this transformation involved much more than basic telephony. The advent of computers and information processing systems brought a new need for a distributed control system; mainframes and centralized information servers needed to be accessed from a multiplicity of points. Some of those points were very far away from the main computer, a world away as the case may be.

By the 1950s and 1960s, engineers began to speak of electronic data interchange (EDI). EDI is a catchall term for the exchange of information through a telecommunications network. The first steps in this direction were taken tentatively, mostly by global corporate managers who realized that access to information meant better business opportunities. But such electronic communications services were seen as having a limited impact in a limited market.

Times have dramatically changed since that prediction was made. The price of computer power has been decreasing dramatically over the past few decades; one dollar's worth of quality adjusted computing power in 1970 cost $73.60 in 1950, only 5 cents in 1984 and less than one tenth of a cent today (Egan and Wildman, 1994). The result has been an explosion of personal computers and a diversified market for software products.

Developments in telecommunications technology represent the next great phase for the development of information products and services. Moving data from one place to another became cheaper and easier because of new communication technology. As transmission systems moved from analog to digital with the advent of the computer, precision and clarity in transmission improved. Circuit and packet switching improved the capacity of the telecommunications network, allowing for its use to be distributed over the entire network rather than simply opening and closing single connections (Wright, 1993). Microwave, satellite, and fiber optics developed and began to improve the opportunities to construct and operate corporate communication networks (Hart, 1988). Instead of thinking of computers as individual machines standing on individual desktops, companies are beginning to think of the telecommunications network as a large computer infrastructure, able to provide distributing computing solutions to seamlessly connect users.

The combination of technological opportunities changed the nature of corporate communications systems but, more significant for our purposes, it altered the way businesses related to the public and private telephone administrations that had provided them with their telephones. And setting up such a global network is not an easy task. Myriad local restrictions affects the kind of technology that can be used and the nature of interconnection between the public and private networks.

But there were companies that saw the opportunity to make it easier. Major corporations began to buy their own equipment from a wide range of equipment providers, and then demand the ability to interconnect that equipment with the public network in each country. By creating their own data networks, corporations bypassed the local and national telecommunications networks and provided for themselves a higher quality of communications services at lower prices.[2]

By the early 1980s, the technological facts were evident. Regulators throughout the world attempted to catch up with the de facto change in the industries by relying on new kinds of service classifications. The idea of "value-added networks" (VANs) emerged as a description of all kinds of communications that extended the use of traditional telecommunications technology to purposes beyond basic telephony.[3] Competition in value added networks became a driving force in the United States, while various PTTs in Europe established value added and public data networks with varying degrees of success.

With the increasing diversification of telecommunications products and services, national and international communications standards-setting bodies came under increasing pressure from the corporations who wanted to buy services and the companies that wished to provide them (Wallenstein, 1990). In the United States and Europe, the response was to propose an open systems regulatory framework, embodied in the Open Network Architecture (ONA) proposal articulated by the Federal Communications Commission and the Open Network Provision (ONP) advocated by the European Union in the late 1980s. In many ways, these regulatory responses were an attempt to accommodate new technologies within the context of existing institutions. But the incredible diversity of possible transmission techniques and equipment types, mixed with the phenomenal increase in micro technology and computer power, has made complete standardization of all products and services almost impossible.

In the developing countries, many analysts became more concerned that the high levels of telecommunications investment on the part of the multinational corporations further isolated them from the citizens of the countries in which they operated. Dependency theorists certainly did raise a valid issue: Much of the global telecommunications and information flow originated in the developed world.[4] Although multinational corporations did bring in important investment dollars, their ability to isolate their networks from the public telecommunications network meant that their investment did not need to contribute to the telecommunications development of the host country.

The result has been a "mixture of old and new structures in international telecommunications," with increased traffic flows between developed countries and between developed and developing countries according to various geographical patterns (Kellerman, 1990). Instead of the unified, established public networks that dominated the first 100 years of electronic communication, a network of networks has developed, some connected, some disconnected, but all dramatically altering our ability to communicate.

These factors have led to the decline of the public network, encapsulated nicely in Eli Noam's phrase used at the beginning of this section. The tragedy of the public network is that all users who have no other place to go are forced onto what is perceived as the lowest standard denominator, the public switched telephone network. High capacity users whose contributions could fuel the refitting of the network with new technologies have opted out in favor of building their own.

The same can also be said about the global information infrastructure, which was linked together by the established coordination of international agencies such as the International Telecommunication Union. As the postal unions before it, the ITU was able to assist in the establishment of a clearing house for international calls, ensuring interconnection and global communications through the public networks. As individual networks are threatened with decline, the fragile set of reciprocal arrangements that have undergirded the established system of global interconnection is also threatened.

New Realities: Changes in the Institutions of the
Telecommunications Sector

The public network as constructed by the public utilities and PTT administrations throughout the world is no longer viable. Regulators and industry participants are well aware of that reality, and structural changes within the industry reflect the technological and economic facts. Taking a brief glance at how telecommunications corporate revenue has changed in the past decade tells a great deal about how these institutions have tried to adapt to changes in business conditions.

In 1980, Public Telecommunications Operators (PTOs) received 53% of their revenue from call charges, which are plain old telephone services. By 1991, with revenues up by over 100 billion dollars in that period, the percentage of revenue from call charges had dropped to 45% (ITU, 1994). What portions of revenue increased during that period? The International Telecommunication Union characterizes it as "other services," which increased from 10% of revenue to 27%. From calling cards to diversified data management products, telephone operators do much more now than simply hook up your telephone. At the same time, the market for telecommunications equipment has boomed from a $20 billion industry in 1984 to a $50 billion industry in 1991.

Overall, though, revenue growth for the largest telecommunications providers has lagged compared to those in other portions of the information and communications sectors. Table 2.1 tells part of that story. The revenue base for many of the large providers has stabilized over the past few years. The big providers in the United States, United Kingdom, and Japan have seen their revenue streams flatten; some, such as NYNEX in the United States, actually posted declines. This is compared to companies and the computer and information services industry that have consistently posed revenue gains in the double and triple digits over the past decade.

The developing world looks different, largely because it is starting from a different place. Countries such as China are experiencing rapid growth in the number of lines installed, showing the strategic emphasis that is being placed on the telecommunications sector throughout the developing world. It would appear, at least on the surface, that the existing technological infrastructure will be the basis of future services in developed world countries, but that the newest, and in some cases, the best technology is rapidly being put into place in developing world countries.

Table 2.1:
Top 20 telecommunications operators (1994 to 1995)

Rank	Operator	Country	Telecom Revenue ($M)	Change 1993-94 (%)
1	NTT	Japan	60,134.7	2.8
2	AT&T	US	43,425.0	8.9
3	Deutsche Telekom	Germany	37,712.6	3.7
4	France Telecom	France	23,288.4	1.8
5	BT	UK	21,262.6	1.6
6	Telecom Italia	Italy	18,047.2	—
7	GTE	US	17,363.0	0.6
8	BellSouth	US	16,844.5	6.1
9	Bell Atlantic	US	13,791.4	10.0
10	MCI	US	13,338.0	11.9
11	NYNEX	US	13,306.0	-0.8
12	Sprint	US	12,661.8	11.4
13	Ameritech	US	12.569.0	5.9
14	SBC	US	11,618.0	8.7
15	US West	US	10,953.0	5.6
16	Telstra	Australia	9,769.3	5.6
17	Telefonica	Spain	9,581.6	5.2
18	Pacific Telesis	US	9,235.0	-0.1
19	Telmex	Mexico	8,655.5	30.0
20	Telebrás	Brazil	7,767.8	-8.9

Based on public information sources, as well as International Telecommunications Union's *Telecommunications Development Report* (1995) and Financial Times' *Telecom Markets* publication (September, 1995).

The difference in employment patterns in the developing and developed world is also evident—almost all of the developed world companies listed in table 2.1 have dramatically reduced the number of employees in the past few years, with BT taking the lead during the 1991-1992 period by reducing staff by 18.9% (ITU, 1994).

Changes in business conditions are paralleled by the fast-paced evolution of telecommunications technologies. The hallmark of those changes has been the transition from analog transmission systems to digital forms of electronic exchange (Neuman, 1991). Any kind of communication can be reduced to digital code, which is basically a string of 1s and 0s that acts as a common language in computer systems. The digital revolution has brought all kinds of communication together, fusing voice, video, and sound into what is now commonly being called multimedia communications.

Digital communication allows for increased speed, decreased price, higher degrees of accuracy, and more capacity on the communications networks. As already discussed, the first stage of the transition facilitated the evolution of electronic data interchange, but that is only the beginning. The hardware of telecommunications is changing dramatically as well. The traditional telephone networks were based on copper wiring and waveform transmission; the infrastructure has remained largely unchanged since the Bell System and the PTT's began to create telephone networks in their respective countries. The capacity of those networks has also not changed much since that time.

The capacity of a telecommunications channel, which is the amount of information carried, is gauged in terms of bits per second (BPS), or bandwidth. The larger the bandwidth, the greater the capacity of the communications channel, the greater the number of bits per second, and, to the thinking of most technologists, the better the network. The average modem or fax machine today works at about 9,600 or 28,800 bits per second. That is also the bandwidth for an average phone line. To have an interactive, full motion video conversation between two people at two different points of the network, you need a throughput of about 45,000,000 bits per second (shorthand is 45 MBPS).

Needless to say, the common phone line is a far cry from the two-way video conferences that AT&T always throws into its commercials. New transmission technologies make such conjectures seem much more than a marketing ploy, however. Fiber optic technology has radically changed the capacity of the telecommunications network, with transmission speeds of up to 100 gigabits per second now possible through the development of Asynchronous Transfer Mode (ATM).

For the most part, fiber optics have been consigned to what is called in the industry trunk and backhaul operations, carrying large chunks of transmissions from one place to another. There is talk of bringing fiber optic technology directly into the home, but the digital revolution has also provided a range of new techniques for the compression of data. By making the size of transmissions smaller, compression allows telecommunications providers to use existing infrastructure, such as copper or coaxial cables, more efficiently.

Whether through compression techniques or the deployment of fiber optic technology, it is clear that the bandwidth used for transmission is becoming more and more plentiful. When we combine that development with the explosions in the computer industry, the multifaceted character of the change in the telecommunications business is even more evident—even though it might not yet appear in the revenue numbers and the investment patterns.

But if revenues are not increasing, and employment is being scaled back rather than scaled up in the developed world, the industry transition underway today is less an expansion than a structural transformation. The kinds of products and services that are conceived of for this new capacity and functionality have yet to be deployed or priced. In most cases, they are more a matter of conjecture than fact. For that reason, a wide spread of assessments exist about the future revenues on the "information superhighway." At the same time, a great amount of cynicism exists about the potential contribution that many of the existing companies will play in constructing and profiting from new technological breakthroughs. In November 1994, a number of Bell Atlantic employees were sent home from their jobs because they decided to wear to work a shirt that depicted Bell Atlantic employees as "road kill on the information superhighway."

Opening the Future: Privatization and Regulatory Change

The common frustration among employees, management, and the technological visionaries promoting the transformation of the telecommunications industry is the perceived inability of regulators and lawyers to keep up with the pace of technological change. Even if those complaints are justified, no one can deny the scope and scale of the dramatic alteration of regulatory regimes throughout the world in the wake of privatization and increased market competition.

All over the world, bypass has driven network operators and regulators to open the market formally to certain kinds of competition. In the United States, the divestiture of AT&T opened the market for competition in two important areas: customer premises equipment and long-distance services. In Japan, both local and international service competition started as Nippon Telephone and Telegraph (NTT) was restructured to face the new economic realities. Great Britain opened its local exchange to a second competitor, and allowed for a range of value added networks that could provide specialized voice and data services (Gillick, 1991).

Developing countries have followed the lead, liberalizing and privatizing their networks to various degrees. In countries where privatization has not occurred, there is a tension between those who do not want to make a huge public policy blunder and others who feel that a continued lack of action is the biggest blunder of all. The countries in the following list have privatized their telecommunications providers in the past decade and a half.

Table 2.2:
Partial list of privatizations in the
telecommunications sector (1983 to 1997)

Country	Company	% sold	date	$US mil.
Argentina	Telefónica Argentina	69	1990	482
Argentina	Telecom Argentina	60	1990	462
Boliva	ENTEL	50	1995	610
Canada	Teleglobe	100	1987	369
Chile	ENTEL	100	1988	N/A
Greece	OTE	6	1996	377
Hungary	MATAV	67	1995	1750
Japan	NTT	13	1986	13850
Republic of Korea	Korea Telecom	2	1993	200
Malaysia	Telekom Malaysia Berhad	24	1990	2350
Mexico	TELMEX	20	1990	1757
New Zealand	Telecom Corporation of New Zealand Ltd.	100	1990	2500
Peru	Telefonica del Peru	100	1996	1350
Puerto Rico	Telefonica Larga Distincia de Puerto Rico	80	1992	142
Singapore	Singapore Telecom	11	1993	2500
U.K.	Cable & Wireless	49	1981	452
U.K.	BT	51	1984	5187
Venezuela	CANTV	40	1991	1900

For developing countries, the traditional route to privatization has been to combine larger telecommunications companies and financial institutions with local interests into investment consortia. The developed world's telecommunications companies have taken large positions as an investment in the huge potential growth of the telecommunications industry in each of these countries. In the meantime, the governments of each of the countries have placed much of the responsibility for improving the management and operation of these newly privatized telecommunications companies on the experienced telecommunications providers.

These concrete policy decisions have radically altered the ownership structures in the telecommunications industry. As pointed out in the previous chapter, the shift in ownership structures brings with it a shift in ostensive goals; managers must be responsive to shareholders and return shareholder value. Some of those shifts are revealed in the investment patterns throughout each of the developed countries. The following charts indicates that privatization has had a dramatic effect on investment in the telecommunications networks of a number of countries.

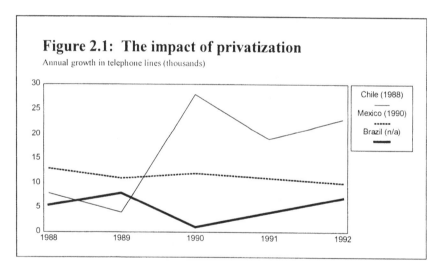

Figure 2.1: The impact of privatization

Annual growth in telephone lines (thousands)

Chile (1988)
Mexico (1990)
Brazil (n/a)

Figure 2.1 compares three Latin American countries and the investment in telephone lines from 1988 to 1992. Two of those countries, Chile, and Mexico completed the privatization of their telecommunications networks by the end of the 1980's. Brazil has yet to privatize Telebrás, their state-run telecommunications provider. The chart indicates that increases in investment occurred subsequent to two of the three privatizations indicated on the chart. Significant increases in Chile are evident; the Mexican experience seems contrary to the experience of Chile and other countries, such as Argentina. Brazil has seen its investment in telephone lines stagnate during the course of that period. This is also true in terms of absolute numbers of lines installed during the same period. Until the mid to late 1980's, all three experience relatively constant rates of line installation. Proportionately, the

Chilean and Mexican telephone companies begin to install a much greater number of lines in 1990, 1991 and 1992 than in the past; the Brazilian investment pattern remains erratic (International Telecommunications Union, 1994).

The press for new investment after privatization has largely been driven by the shareholders and government regulators, both of which hope that an expansion of the network's capacity will lead to an increase in traffic and revenues. Privatization, in other words, has brought in an influx of foreign currency like none that the developing world has ever seen, and privatization of telecommunications networks has been at the forefront of these developments.

Nevertheless, as has been discussed in the earlier chapters, privatization does not necessarily change business practices. Business practices are connected to regulations, and the regulatory environment that defines the provision of services in these countries has not changed dramatically since privatization. As we will see, the levels of competition are restricted by political and economic decisions that run contrary to the principles of liberalization. Energies have been released over the past 3 to 5 years that will now need to be directed, especially as the flow of money to the developing world slows dramatically in reaction to recent currency problems worldwide.

Interim Conclusions

If there is any one word that would sum it up the discussion of the transformation of the telecommunications sector, it is diversity. The global result of technological and economic changes is a hodgepodge of public and private networks and a daunting diversity of public regulations and technical standards.

The deeper dynamics driving the revolution in telecommunications are attached to broader economic trends; we face a technological revolution that appears to be pushing the networks of the future to further and further decentralization, yet we come from an age where the political economy of our media and the mass psychology of nation states and national information leave strong centers that continue to attract social forces (Neuman, 1991).

What we see in the history of the telecommunications sector to date is a set of political decisions that have determined the provision of service. It was a political decision to establish the public utility model for telecommunications development and it was a political decision to attempt to sustain it in the face of the changes described. There is an opportunity to make political decisions today that will determine the future economic and technological course; the hard technological determinism that many have stated will guide the future of the industry does not stand in light of the possible pasts and futures that lie before us.

The old arrangement will not do; public management and monopoly arrangements are no longer tenable in the new technological environment. Nevertheless, the necessities of transitions and the realities of bureaucratic interests force us to recognize a critical reality: Public and private management will have to coexist for a while, if not indefinitely. During that period of difficult coexistence, institutions will transform themselves to reflect the new technological and social opportunities. Until then, the telecommunications sector is in desperate need of ways to bridge the public and corporate management of resources.

LINKING TELECOMMUNICATIONS
TO DEVELOPMENT

One possible bridge is the framework provided by the broader theories of social and economic development discussed in the first chapter. And telecommunications, by its very nature, lends itself well to analysis by development theory by the nature of its definition and role in a community or society.

Perhaps the best definition of telecommunications comes straight from its roots. It is communications through a medium. The word communications, defined as "a process by which information is exchanged between individuals through a common set of symbols, signs or behavior," comes from the word community. The whole idea of telecommunications is closely tied to how communities work, which is why so much has been written about the impact of new technologies on political and social organization (Abramson, Arterton and Orren, 1988; Dutton, Blumler, and Kramer, eds., 1991).

Telecommunications is the critical technology of the information age. In a world that increasingly depends on the integration of information and the automation of control systems that coordinate the activities of so many machines, the technologies that connect us define the way we work, live and learn. For that reason, the telecommunications revolution has become a global phenomenon. Talk of establishing a Global Information Infrastructure is a dominant theme in the literature, and many speak of the potential value of a world wired for information.

The social implications of the telecommunications revolution are far-reaching. In the first chapter, we alluded to the relationships drawn between social and technological change. For the study of telecommunications development, this relationship remains a central concern. There is a real connection between the kind of technology that is used to communicate and the kind of society that exists with the technology. As technology changes, what a person can do changes, and how people relate to each other also changes. Development is concerned with what a person can or cannot do, to re-echo the words of Amartya Sen, and changes in telecommunications technology clearly change what a person can or cannot do.

A number of questions immediately arise: Does the expansion of the telecommunications network help a community to achieve the goals of development? What kinds of policies, aid packages, regulatory stances, and corporate management decisions would most appropriately tie telecommunications development to national development? What policies focus investment on the expanded opportunities for modernization and participation associated with new telecommunications technologies? And, most importantly for our purposes: Are there specific kinds of technology and technological organization in the telecommunications industry that contribute to the development process?

Were the telecommunications sector less significant to the needs of national and community development, these questions might not be so difficult to unravel. In applying the needs of development to the telecommunications sector, a significant relationship is revealed that can not be ignored by policy makers or corporate strategists. Both have a vested interest in the goals of development in the broadest sense of the term—not just the development of the telecommunications sector.

Correlations and Causal Relationships

Telecommunications development is traditionally connected to national and economic development through correlations and analysis of causal relationships. The most significant correlation, which shows a relationship between income per capita and penetration of telecommunications services, is depicted in Figure 2.2.

The data used for this regression analysis come from the 1994 World Telecommunications Development Report, published by the International Telecommunication Union. On the vertical axis of the regression analysis is the Gross National Product per capita of 135 countries throughout the world. The lowest in the sample are below $200 per capita, and the highest are over $20,000 per capita on the horizontal axis. As the statistical analysis shows, there is a close correlation between the penetration of telephones in any given country and the per capita income of that country. This relationship has been shown many times in research, and the most recent numbers from the World Bank and the International Telecommunication Union show no real change. Generally speaking, the higher the income per capita is, the greater the penetration of telephones in the country (Katz, 1988). "In broad terms," according to one economist, "each $1,000 increase in per capita GDP tends to be associated with an extra three lines per 100 people" (Wheatley, 1996).

It is possible to layer regression analysis over regression analysis; there are definitely correlations between increased penetration of telecommunications services and decreases in infant mortality, higher levels of education, penetration of mass media, and other traditional measures of development. This shows only that the road to development is somehow paved with phones. What it does not show is a causal link between investment in telecommunications and the achievement of development goals, such as modernization or increased social participation (Brynjolfsson and Hitt, 1993; Strassmann, 1994).

Such causal relationships are often asserted. Ithiel de Sola Pool's book on the social uses of the telephone, written in 1977, holds many such claims. Perhaps the most famous is that the two-way communication of the telephone is a participatory technology, while the passive, unidirectional information flow of the television does not allow for participation (Pool, 1977). It is likely that we could plot a regression analysis to show the relationship between participatory democracy and telephone penetration as well, but that would still lack for a causal relationship.

Figure 2.2:
Correlating GNP and penetration of basic telephony.

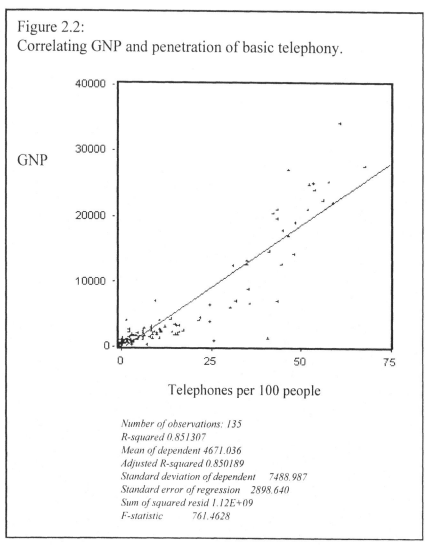

Telephones per 100 people

Number of observations: 135
R-squared 0.851307
Mean of dependent 4671.036
Adjusted R-squared 0.850189
Standard deviation of dependent   7488.987
Standard error of regression   2898.640
Sum of squared resid 1.12E+09
F-statistic   761.4628

The most significant relationship between telecommunications and development has been drawn through productivity. Telecommunications makes labor and capital more productive; productivity spurs economic growth and development. Telecommunications therefore promotes economic and social development. This sounds logical enough, but finding the data and analysis to definitively show such a relationship is a difficult enterprise at best. Perhaps the most detailed, methodologically rigorous expression of this relationship comes from Francis Cronin of DRI/McGraw Hill.

One of his most substantial works on this subject was the Pennsylvania Telecommunications Infrastructure Study. Cronin's research was funded, in part, by a consortium of Pennsylvania telephone companies and was conducted in conjunction with Deloitte and Touche. The final report, issued in March of 1993, and Cronin's subsequent publications will guide this discussion of the relationship between telecommunications, productivity, and economic growth (Cronin et al., 1993).

The key to economic growth in the United States has been total-factor productivity, and this becomes the starting point for the analysis. From the period of 1889 to 1988, total-factor productivity increased by 1.6% a year, accounting for over 50% of the total increase in GNP for that period of time. When Cronin disaggregated total-factor productivity gains, he reveals certain sectors have contributed to the improvements in productivity while others have done less. Telecommunications has outpaced the economy in total productivity growth, "implying not only that it has contributed its share of total output more efficiently, but that it has contributed to overall productivity growth. Conversely, without advances in telecommunications production, the U.S. economy would have experienced greater declines during the 1970's and a slower recovery in the 1980s" (Pennsylvania Public Utilities Commission, 1993).

In analyzing the increases and decreases in factor productivity within other industries, he discovers that the use of telecommunications as a production input has increased markedly from the period of 1965 to 1987. Key services critical to the modern economic structure, such as finance and insurance, have seen their use of telecommunications services increase by over 2.9% a year. In total, through a direct increase in the productivity of the telecommunications sector and its contributions to the increased productivity of other sectors, the research indicates that telecommunications has been responsible for 15% to 35% of the total economy-wide productivity gains.

What this research does is give a more direct linkage between economic development and telecommunications investment through increases in total factor productivity. Even so, "we must recognize that telecommunications is only one of several inputs necessary for economic development," and, although Cronin's study does suggest a strongly positive relationship between economic growth and telecommunications investment, the specific examples outlined in this discussion are focused on one state in a highly developed country.

The Response to Causal and Correlative Relationships: The Investment Models

The causal links do not suggest specific courses of policy beyond what should be clearly evident: Telecommunications is tied to development, and investment in telecommunications has contributed to economic growth in the United States, and attention needs to be paid to the telecommunications sector if we are to shape the development of a community. If we can get phones and communications technology into the hands of citizens of every class and community, we are likely to have a positive impact on economic growth.

But, to date, most of the models for discussing telecommunications development in light of the broader needs of social and economic change has been centered around the activities of nation-state bureaucracies and international institutions. That kind of bias is clear in the 1994 *World Telecommunications Development Report*. It lists six telecommunications development constraints.

- Lack of investment;

- Foreign exchange scarcity;

- Investment inefficiencies;

- Organizational limitations;

- Inadequate private-sector involvement; and,

- Insufficient regional co-operation.

Each of these problems are certainly significant. Developing countries spent less than 20% of the almost $120 billion invested in telecommunications worldwide. Sometimes operating profits from telecommunications providers in developing countries are no reinvested into the network but are siphoned off to subsidize government programs. In other cases, there is a lack of money, pure and simple. A lack of foreign exchange restricts the opportunity to import telecommunications equipment in order to keep a network functioning. The costs of running a network and installing a line are much higher in developing countries, in some cases the average cost of installing a line is over $20,000 per person (as compared to the often quoted figure of $1,500 in the U.S.). Human resources are limited, there is no private sector to speak of, and the potential for using regional resources is never explored, much less exploited.

Each of these issues is construed by the writers of the World Development Report as being more applicable to the developing world than the developed world. But, in many ways, each of these pitfalls is applicable to telecommunications planning in the developed world as well; without appropriate levels of investment and organizational capacity, the telecommunications sector will not flourish and the goals of economic and political development will be affected.

The traditional answer to meeting these pitfalls is a combination of better planning and multilateral assistance. On the management side of the equation, the high performer countries have been able to ensure that profits generated from the telecommunications provider have been reinvested into the operations of the network and that a consistent ownership and regulatory environment has been maintained.

The main agencies for international telecommunications assistance are the International Telecommunication Union and the United Nations Development Program, which have provided a range of targeted programs to enhance the telecommunications sector. The World Bank and European Bank for Reconstruction and Development (EBRD) have committed an increasingly greater percentage of their resources for loans to the telecommunications sector; the restructuring loans and assistance with privatization efforts worldwide have contributed greatly to the recent explosion of activity in the telecommunications sector.

Regional cooperation is also seen as a critical element of telecommunications development worldwide. A number of regional economic and political associations have begun to take an active interest in the development of the telecommunications sector within their region. Some of the main intergovernmental telecommunications organizations include the Asia-Pacific Telecommunity, the Pan-African Telecommunications Union, the Conference of European Postal and Telecommunications Administrations and the Inter-American Telecommunications Commission.

Consistent with all of these approaches has been a reliance on large-scale organizations to push the development of the telecommunications sector in each country. Either through a multilateral agency or through a public utility such as a state-owned telecommunications provider, it has been assumed that telecommunications development depends on centralized institutions and forces. That raises an interesting question: Is there a more appropriate institution or kind of institution for the telecommunications industry that does not rely on scale and scope to sustain itself? Are there institutions that can evolve from a limited subscriber base, even in the most competitive of environments?

The answer to that question clearly depends on how technology is introduced into new markets and environments. If we knock on the door of a social scientist and ask for an appropriate model for us to consider, the most likely response would be for the social scientist to draw a diffusion curve on the blackboard. Everette Rodgers (1962) provided the classic study of the diffusion of information and new practices in a social environment—and the basic output of his analysis was an s-shaped curve. When a new practice, piece of information or technology is introduced, few adopt it initially. Soon, it picks up momentum, and levels of penetration increase substantially for a short period of time as the mass public begins to adopt the practice or receive the new piece of information. Finally, as the saturation point is reached, the rate of penetration decreases to zero.

A variety of factors influence the diffusion of technology at each stage along this curve. The introduction of a technology can fail if it does not meet a perceived social or political need. After the introduction of a technology, certain structures have to exist to facilitate the diffusion of the technology, such as its adoption by leadership groups or its utility relative to other kinds of goods and services available. Finally, the transformation of the social or economic organization much be such as to sustain the continuing use of the technology.

In an environment of limited competition, the institutions responsible for pushing new products and services up along the diffusion curve are easily identifiable: the large scale efforts of national states, dominant providers and international lending organizations. But, in a more competitive environment, individual responsibilities for this task are somewhat more difficult to identify. There are situations where service providers stall the diffusion of technology for competitive advantage, or governments overregulate new technologies or services for purposes of tax revenues or protecting markets. An informed strategy to facilitate the introduction and diffusion of technology has to be informed by issues of political economy along with the cost benefit of corporate policy.

That is why, in a competitive market, policy needs to be concerned with diffusion of new and existing services. In turn, policy institutions need to define the ground on which new technological investment and diffusion are of commercial advantage to telecommunications providers. But this will not be as easy as it once was; competition, by its very nature, should reduce the amount of leverage that public sector institutions have over private sector institutions. Sound arguments which speak to the interests of corporate institutions will have to be presented, rather than dictated, to competitive service providers.

Therefore, the first part of the value proposition which links public and private sector institutions in the development process needs to be a new conception of how technologies and services can be driven and diffused into new markets. But then we knock on the door of the market researcher, and a different sort of model appears on the whiteboard in the researcher's corporate office.

A market researcher would focus on a traditional demand function as it relates to the setting of prices for a particular good or service. The penetration of the good or service would be depicted on the horizontal axis, while the price of that good or service would appear on the vertical axis. For telecommunications services, as is the case for almost any product made available in a market economy, there is a price at which no person would purchase the good or service. There is also a price at which every person would choose to purchase the service. Between those two points, there is an appropriate demand function that charts the quantity demanded at any given price.

For telecommunications services, we speak of similar relationships: If I were to give you a cellular phone for free, to use as much as you want for free with, as it were, no strings attached, you would probably take it. If I asked you to pay upwards of a million dollars for it, you would likely decline. In between those two points, there are different price levels you might accept depending on the utility you saw in having a cellular phone.

This is basic microeconomics, but it sets the stage for an important element of the discussion. The first chapter argued that competition requires the existence of a number of microeconomically sustainable institutions that can drive the penetration of services. As technologies diffuse through a society, a certain amount of time will be required for the momentum to pick up. This is especially true for lower cost technologies, where margins will be thinner and new participants are less likely to have a buffer to prevent collapse during economic downturns.

To combine sustainability with a strong program to diffuse new and existing services throughout markets served by competitive providers, therefore, requires an examination of new competitive institutions in the telecommunications sector. But particular attention needs to be placed on the structure and purpose of those institutions, and how their organization might be shifted to capture the benefits associated with economic and social development.

NEW INSTITUTIONS, OLD PROBLEMS:
THE NEEDS OF DEVELOPMENT IN A COMPETITIVE ERA

The problem is that there are few examples of what a truly competitive telecommunications provider looks like—the history of competition in the telecommunications sector has been short indeed, and, for most countries around the world, the first page of the first chapter has yet to be written. Nevertheless, there are places to look for direction in this area. If the determination to establish the public utility structure for telecommunications services was, in part, a political decision, and if alternatives to that kind of technological and corporate organization are available, it should be possible to theorize about the structure of alternative institutions for the provision of services.

The implications of Milton Mueller's argument on the development of competition in the American telecommunications market now comes into full relief. Access competition, which is his description of the early push for telephone penetration in the United States, stands out as an interesting alternative for telecommunications development:

> The historical facts about access competition have important policy implications. If the standard historical assumption about regulated monopoly's role in the creation of universal service is true, then developing countries should stay with regulated monopolies to develop their infrastructure before experimenting with competition. If, on the other hand, access competition played a critical role in the developmental stages of the US infrastructure and this experience accounts for the tremendous US lead in the extension of telecommunications service, then a very different policy conclusion can be drawn.

> Conditions in developing countries, which have low penetration and a stagnant monopoly, often closely correspond to the conditions in the USA prior to independent competition. A policy of open entry and systems competition could have similar effects, although of course there are many differences in conditions. (Mueller, 1993, p. 369)

Neither the technology nor the history are the same today. Nevertheless, it might be possible to establish and sustain access competition using a different set of technologies and corporate institutions, thus unlocking the energies of the telecommunications sector. There is a microeconomic and micropolitical level of analysis embedded in this proposition, both of which require explanation and definition.

Increasing penetration for telecommunications services depends on reducing the cost of the service. Reducing the cost of the service requires reducing the costs of one or more of the inputs of that service, thus shifting the demand curve downward and driving higher rates of usage. At the same time, institutions must be constructed so that they can competitively coexist on a level playing field. That is the sticky issue in telecommunications services today. The fixed costs of providing telecommunications services have been extraordinarily high, from the standpoint of both human resources and technological applications.

Creating competition requires us to rethink the whole range of cost inputs so as to understand how costs might be lowered through the selective application of new technological applications. That becomes the critical microeconomic criterion that will spur telecommunications development through competition.

The politics are even more problematic. Regulatory bodies can be described as both reactors to and shapers of technological and organizational processes. In examining the taxonomy of regulatory agencies throughout the world, we see that they have been more reactive than proactive when it comes to the application of new technology. The political decisions since the turn of the century have largely been made by default, with monopoly interests dominant in virtually every country throughout the world.

The political community should not protect new institutions and technologies as they enter onto the scene for reasons of competitive advantage; rather, regulatory institutions can choose to foster competition through identifying the opportunities for new entrants in a competitive marketplace. By assessing technologies and suggesting kinds of technological and organizational institutions appropriate to competition, the policy community and regulatory institutions can perform a vital public service: They can point the direction to a competitive marketplace and can help to lead corporate institutions in a variety of appropriate directions.

The results will be an accelerated pace of development in the telecommunications sector. As the U.S. Department of Commerce's 1991 report, *Telecommunications 2000,* argues, "competition builds markets; it does not erode them. The more competitive a given geographic, product, or national market, the faster it seems to grow. Competition thus seems key to sector development" (U.S. Department of Commerce, 1991). The accelerated rate of development through competition will, in turn, enhance the opportunities for participation and modernization.

The goal is, therefore, to identify specific technological and organizational characteristics appropriate to sustainable competition in the telecommunications industry. As a starting point, we should look to existing forms of competition in the telecommunications industry and focus on one particular example: competition for wireless access to communications services.

A Taxonomy of Competition in the Telecommunications Industry

Three specific markets are generally described as competitive in today's telecommunications sector. Present arguments, though, are largely mired in the politics of deregulation. This is especially true in the U.S., where virtually every corporate institution has a hired academic or two who can effectively argue that the market the company dominates is competitive while other markets the company wishes to enter is not competitive. How competitive each market is, therefore, provides an opportunity for much debate, but the structure of all three is generally based upon the establishment of at least the facade of competition, if not the reality.

The first kind of competition has already been discussed in some detail: Specialized data and voice networks for corporate users. As the interface between the telecommunications and computer industries, data networking is cluttered with all kinds of network providers, hardware producers, consultants, and software makers. Clearly, a great deal of competition exists for this kind of telecommunications service.

The second kind of competition most often discussed is competition for long-distance services. The development of competition in this sector depended on a certain kind of technology: point-to-point microwave transmission, which became commercially viable in the mid to late 1960s. Among the first companies to use this technology was Microwave Communications Inc., which later shortened its name to MCI.

MCI began to compete with the long lines division of the then monopoly provider in the United States, AT&T. After years of political maneuvering, Judge Harold Greene declared that long-distance services could sustain competition if the "local bottleneck" could be separated from the long-distance network. AT&T, under an agreement put together by the company and the Justice Department, agreed to divest itself of the local exchange in order to compete for long-distance service and equipment contracts. The resulting arrangement left the Regional Bell Operating Companies as regulated monopolies responsible for the local exchange, while MCI was granted the access to the existing wireline telephone system to compete for long-distance services.

Soon after they achieved that goal, MCI stopped using its microwave towers (Huber et al., 1993). The battle for market share in long distance would not be fought in the United States with microwaves, but rather with fiber optic technology. The huge capacity of fiber optics allowed long-distance carriers to roll-out vast networks that connected local service area with local service area. Peter Huber has argued that the resulting introduction of fiber in the long-distance services has resulted in a noncompetitive situation. His argument was based on the huge overcapacity in the network and the fact that shares in long-distance services in the United States have stabilized in the past few years. Prices were pushed down initially, but that does not mean a competitive situation existed or currently exists.[5]

Similar kinds of wireline competitive markets have evolved under the aegis of evolving regulatory liberalization in a number of developed countries. During the 1980s, the British Department of Trade and Industry (DTI) helped to bring into existence a competitive duopoly for wireline services; BT faces stiff competition in long-distance and business networks from Cable & Wireless' subsidiaries throughout the country. The market still largely belongs to BT, even though new competition might change that fact dramatically in the years to come.

NTT of Japan faces a similar challenge. The Japanese government has chosen to license a number of service carriers to provide for competition in local, long-distance and international services. Slowly but surely, NTT has had to face additional entrants into the market for a wide range of telecommunications services, in particular long-distance.

The model of long-distance competition and local monopolies has been replicated in a number of developed and developing countries throughout the world. As part of Mexico's liberalization scheme, long-distance markets will be opened to competition in the next few years, whereas local telephony is scheduled to remain under the monopoly control of TELMEX until the year 2037.

It is important to point out that, at the foundation of the initial competitive markets for communications services, the access technology was wireless. Wireless presented certain advantages for establishing a low cost network that allowed MCI to argue that they should be allowed into the market. Wireless required less network maintenance, because operators did not have to worry about all of the wires in the network. The control organization for a point-to-point network was easier to manage than a distributed wireline network, although complexities in switching were certainly just as complex, if not more.

So why the transition to fiber? Because of the high capacities of fiber optics, and the ability to transmit digital information instead of analog waveforms, the quality of service could be improved. But the result has been a non-competitive situation, according to Peter Huber. He argued that the incredible overcapacity in the long-distance networks in the United States has made a farce of competition; AT&T could easily cut its prices, make a profit, and run the competition out of business. Huber called it "Candice-coated competition," echoing Candice Bergen's role as spokesperson for long-distance provider, Sprint.

If that is true, why doesn't AT&T just do it? Even after breaking up into three different companies, AT&T would have more than enough motivation to do so. First, according to Huber, the regulators in the U.S. will not let them. Second, even if they did, the antitrust division of the Justice Department would be knocking at the door and filing a flurry of legal briefs, calling for another divestiture in order to support a competitive marketplace.

We can also add another layer to Huber's argument. If the best opportunities for telecommunications development are based on extended penetration and access, long-distance competition can not help. Long-distance, as defined in the U.S. regulatory model, connects local loop to local loop, not person to person. Is that the kind of competition that will bring to people the new services and access that could generate the concurrent effects of economic growth and increased productivity? Probably not.

The present market for long distance may be competitive, but the true price and service differentiation that would be required of a competitive market has not emerged. That is in great part because differentiated service offerings have yet to emerge; phones are still phones, no matter what the future vision of the telecommunications networks may be. Until there is a sense of differentiated service based on price, content, and access, the long-distance market can not stand as a model for widespread competition in the telecommunications industry.

Wireless technology has also been at the foundation for a third kind of competition in the telecommunications marketplace globally: wireless access. In the United States, the most prevalent form of wireless access has gone under the name of cellular telephony, with mobile phones targeted mostly to high-end customers. In a number of countries throughout the world, multiple providers have been licensed to provide competing services in a given geography. In most cases, that geography has been national, but, in the United States, a variety of large and smaller companies have developed to provide service in a wide range of service areas.

It is important to point out that the character of competition for wireless access in the US and throughout the world has been far from vigorous. A wealth of studies exist to test the character and quality of competition in existing markets in the United States, with most of them echoing a common theme: The amount of real competition is, at present, minimal (Ruiz, 1994). Nevertheless, the market for wireless telephony has sustained a number of providers, brought new entrants into the telecommunications sector, and transformed the possibility for local and long-distance competition.

More importantly, wireless access is the only kind of competition that has reached directly into the local market, to a broad base of consumers and corporate users. On that basis alone, George Calhoun (1992) is right in saying that the advent of broad based opportunities for wireless access means "the end of the natural monopoly." He wrote:

> The introduction of radio technologies into the long-distance segment of the telecommunications market in the 1950's led directly to the break-up of the monopoly for long-distance services. Freed from the cost constraints of wireline plant in the long-distance application, operators found that they could effectively compete, and tolerate competition, without destroying either themselves or the market. It took the regulatory structures some time to catch up; however, competition was eventually embraced, and today we have AT&T arguing that it has come to the point where they should no longer be saddled with the assumption that they are the dominant carrier.

> Wireless access will work the same transformation in the local exchange market. Make no mistake about it, we are witnessing the beginning of the end of the natural monopoly. In the twenty-first century access environment, there will be no reason why there cannot be as much competition in telephony as there is in, say, the airline business or the trucking business. (Calhoun, 1992, p. 121)

The examples of the airline or trucking business are worth an additional comment. An airline or a trucking company owns the facilities it uses to provide services. Both kinds of companies use commonly accessed infrastructure; trucking companies use the same publicly owned and maintained road, and airlines use municipally owned airports. So part of the competitive environment is based on the ability of each company to maintain efficient and profitable facilities, but the other part involves ensuring access to the commonly available infrastructures.

By analogy, the competition in the telecommunications industry which Calhoun described would involve some commonly accessed infrastructure, and some that is owned by the company providing the services. Yet that constrains the kind of competition possible, keeping it a step away from a market where each company owns the facilities required to deliver products to customers. The more commonly accessed infrastructure, the more opportunities for market inefficiencies and political disputes that have less to do with the quality and cost of services and more to do with the political muscle of the institutions vying for access.

In a marketplace with a history of placing regulatory and political needs over customer interests and concerns, this can be a recipe for disaster. For that reason, Calhoun's hope for a market as competitive as the airline industry or the trucking industry is not going far enough. Facilities-based competition should be a long-term goal, turning the information products marketplace into a environment which resembles the competition between Coke and Pepsi, instead of Delta, Lufthansa, JAL and United.

Leaving aside the argument of whether or not the market for wireless communications is presently competitive, wireless communications has been the foundation for what little direct, facilities-based competition presently exists in the telecommunications industry. If we are looking for a mechanism to hasten the penetration of competition and new services in the local exchange, then we need to look to wireless access for some possible answers.

CONCLUSIONS: SUSTAINABLE INSTITUTIONS AND EFFECTIVE COMPETITION

To recreate access competition in this modern era, a platform for competition needs to be identified that would be appropriate for improving local access to communications services. The platform has to be sustainable and provide for facilities-based competition among a variety of service providers. It would also have to be amenable to the needs of national development already outlined in the opening discussion.

Considering the role that wireless technology has played in the development of competition in the provision of telecommunications services worldwide, it is likely that it can be used as a technological foundation to support further competition. We have come to a tentative conclusion about the best possible "strategic" choice for "strategic liberalization."

That brings us back to the possibility of using competition to fuel economic and political development through the telecommunications sector. If the traditional institutions within the telecommunications sector can only go so far in exploiting the new opportunities for development opened by social and economic change, other kinds of technologies and institutional organizations will need to evolve.

The heart of the argument remains ahead. Having established the link between telecommunications and development, identified the critical issues for telecommunications development, and described the existing kinds of competition in the telecommunications sector, there is now a solid base to test the following assumption: It is possible, and desirable, to employ wireless technology as the basis for competitively sustainable corporate institutions. More than that, it is essential to the future development of the global information marketplace. The next chapter expands the discussion of wireless access in preparation for a more detailed description of how the policy of strategic liberalization could be implemented and what kinds of institutional organizations and behaviors might arise in a competitive marketplace.

ENDNOTES

[1] Two of the largest telephone companies in the world, Sprint and GTE, were born of this attempt at realizing universal service. By using these two companies (or, more properly in the case of Sprint, the ancestor of the company) to provide subsidized service to rural communities, the government fostered the penetration of telephony and strove to meet the goal of universal service.

Interestingly enough, that heritage has offered GTE a competitive advantage in the wake of the 1996 Telecommunications Act. The company has been able to forestall implementation of the act in the courts—creating boundaries for the Regional Bell Operating Companies in areas like long-distance while leaving the field open for this "non-monopoly" company to expand its market presence.

[2] A prime example of companies that profited on this technological revolution is EDS. Bought by General Motors to manage their international data network, EDS continues to grow today in a marketplace that has begun to become cluttered with companies that operate data and communications networks for major corporations.

[3] Eli Noam, "International Communications in Transition," from *Changing the Rules: Technological Change, International Competition and Regulation in Communications,* Robert Crandall and Kenneth Flamm, Eds. (Washington, DC: The Brookings Institution, 1989). Noam's definition of VANs reflects the ambiguity of the term. "Value added services are merely a functional category and not a regulatory term," which Noam saw as a necessary transition step from plain old telephone service to the diversity of new products and services that the industry wishes to make available (p. 273).

[4] There are a range of hotly disputed facts and figures that have been used to illustrate this point. In the early 1980s, with the MacBride Commission report to the International Telecommunication Union acting as the catalyst, an extended debate about the effect of this trend reached a climax, although the issue has been left largely unresolved. At the center of the argument about cross border information flows was the "free flow of information" mod-

el espoused by the United States and many other western countries and the belief that developing countries needed to practice a form of communications protectionism in order to prevent the erosion of their national culture. The later argument was most commonly associated with the dependency school of development theory.

The geopolitical change of the past decade has largely altered the terms of the debate, although the analysis buttressing the countervailing points of view are extremely relevant to the discussion of telecommunications and development. See, in particular, Ithiel de Sola Pool, *Technologies of Freedom* and *Technologies Without Boundaries* for arguments for the free flow of information model. The arguments for cultural protectionism are well expressed in Cees Hamelink, *Cultural Autonomy in Global Communications: Planning National Information Policy* (New York: Longman, 1983).

[5]   When writing and speaking in the early 1990s, Huber openly admitted being in the employ of the Regional Bell Operating Companies. The vested interest he had in the outcome of some of the research he conducted is reflected in these argments: his clients had a vested financial interest in making the point that competition did not exist in long-distance, but rather that significant competition was present in the local loop. That argument was used to justify the need for entry of the local providers into long-distance service in the US in the early to mid-1990s.

Wireless Communications: A Foundation for Competition and Development

Wireless access is generically defined as the use of radio to provide access to the telephone (telecommunications) network (Calhoun, 1992). The main enabler is the electromagnetic spectrum, which can be used to carry encoded information from point to point and sustain a conversation between two people, or two computers for that matter.

It is difficult to find a pessimistic projection for the growth of wireless access services—in fact, just looking at the numbers, it would appear that analysts are competing to develop the rosiest possible predictions based on existing growth data. In the United States, for instance, 1995 to 1996 growth in subscribers for wireless access, including cellular and other new services such as Personal Communications, was more than 25%. Donaldson, Lufkin, and Jenrette (1996) predicted that, by the year 2005, more than 131 million people will subscribe to either a cellular or a PCS network in the United States. By 2001, more than 80 million Americans will have pagers, up from just under 20 million at the end of 1993. For the rest of the world, DLJ predicted a total of 293 million wireless subscribers by the year 2000, almost six times as many as at the end of 1995.

The race ahead in bringing subscribers onto the network has dramatically changed the investment patterns of many telecommunications companies. Telecommunications providers throughout the world have responded by dramatically stepping up investments in wireless access networks. In 1995, companies in Norway, Denmark, and Sweden, for instance, were installing between 6 and 7 times more wireless lines than fixed lines. Similar patterns will appear in the chapter on Brazil, where the regional companies that are part of the Telebrás network have been investing more heavily in their new cellular properties than in their wireline networks. In total, if the DLJ estimates turn out to be correct, providers around the world will have invested more than $100 billion dollars in building network infrastructures to support the demand.

Table 3.1:
Wireless Local Loop Subscribers
2000 to 2005 (Millions)

Region	2000	2005
Developing World		
Asia Pacific	18.10	106.50
Latin America	3.80	15.60
Eastern Europe	3.40	11.40
Africa	1.60	9.60
Middle East	1.90	4.60
Developed World		
North America	13.00	21.80
Western Europe	10.20	17.60
Asia Pacific	8.00	13.80
Total	**60.40**	**201.50**

Note: MTA-EMCI, as reported in the Financial Times' *Telecom Markets* publication in May, 1996.

For that reason alone, manufacturers of equipment for wireless networks are particularly confident of the future. Ericsson's company data predicted that almost 450 million people would be subscribers to wireless networks by the year 2001 (a bit higher than the DLJ prediction, it should be noted), with more than 150 million in the Asia Pacific region alone. The company believes that the evolution of the market will be driven by the transformation of services to meet mass market needs, as well as the potential for increased product differentiation between service providers. In addition, Ericsson thinks most subscribers throughout the world will have access to digital networks. The older analog networks will begin to fade towards the end of the century, leaving only the most advanced technology in the hands of the customer.

Then there are fixed wireless systems, often described as wireless local loop (WLL) applications. Some analysts predict that more than 200 million will be served by these kinds of networks by the year 2005. Table 3.1 offers the detailed projections of telecommunications analysts MTA-EMCI, as reported in the Financial Times' Telecom Markets publication in May, 1996.

The greatest growth is projected in the Asia Pacific markets, where literally billions of people are still without access to telephony. Significant growth is also predicted in Latin American and Eastern European markets, where wireline services have not been able to reach even a majority of the population. In developed world markets, where wireline services dominate and cellular telephony has a broadly established market presence, the predictions are for less growth between the years 2000 and 2005. The group predicts that almost half of those served by wireless local loop applications will have access to digital networks. Clearly, this would be a significant, even revolutionary, implication for information and telecommunications networks in both the developing and the developed worlds.

The social consequences of wireless access to telecommunications services have been a topic of discussion for years, and with the white-hot growth of the industry continuing, all sorts of comments have been made about the use of cellular and wireless phones in developed and developing world contexts. Wireless access represents an opportunity to alter the basic patterns of interaction in corporate environments, leading many to conclude that "developing business structures will demand greater use of the radio spectrum" to implement decentralized administrative and bureaucratic structures (Forge, 1996). The difference between business and personal use of cellular telephones in countries like the U.S. has become blurred as people begin to use wireless access to manage their personal and professional lives (Katz, 1996). In the developing world, business communities have been revolutionized by the opportunity to bypass outdated wireline networks in favor of increased mobility and access (ITU, 1994).

What does this all mean for the industry and, more importantly, how do public and private sector institutions take advantage of this opportunity? So far in this book, the argument has been made for using wireless communications as a focal point for increased competition and the sustainable development of the telecommunications sector, in alignment with the principles of liberalization and increased market competition. Although a variety of corporate strategies and public sector programs have been advanced as a response to the incredible growth of wireless communications, many are based on existing service models—and, as this chapter makes clear, the existing service models are not a strong basis for the sustainable development of the industry.

What we are looking for, then, is not just any kind of competition, but competition that meets the goals and objectives outlined in earlier chapters: increased penetration of services, enhanced levels of political and economic participation from customers, and a strong partnership between public and private institutions that participate in the telecommunications industry. In other words, we are looking for models of wireless access that are appropriate candidates for the strategic liberalization of the telecommunications marketplace worldwide. This chapter sets the stage for our search through the four case study countries by providing an overview of existing models for wireless access and a discussion of the challenges for corporate and government institutions face in making strategic liberalization a reality.

THE EVOLUTION OF WIRELESS ACCESS SERVICES

The service provided by telecommunications operators is "access to communications services." Providers can offer access by two means: stringing a copper or fiber-optic wire from one point to another, or by using the electromagnetic spectrum to connect one point to another. So, in the broadest sense, wireless access is defined, for purposes of this discussion, as the direct connection between a person and a telecommunications network by transmitting information through air.

As an idea, wireless access is not new. Briefly mentioned before was Marconi's first applications of his wireless communications system—wireless telegraphy. Two-way, wireless communication has been around for more than a century, so we are not dealing with a conceptual novelty per se. It is the application of the technology that is new, and the application of the technology is enmeshed in new innovations and regulatory changes over the past 100 years.

Wireless access, though, is only part of a whole; the networks of the future will include both wireless and wireline access. And many networks will be able to compete using only wireless access as the foundation for their network infrastructure. Until recently, it was inconceivable to think that the entire public switched telecommunications network could be replaced in any country. The fixed costs of wires made such a conception look more like folly than practical policy. But it is now possible to do just that. An entirely new communications network can be built in any country, even the United States, with wireless access as a critical component of the service provision. Without wireless access, corporate and public managers would be left with the present network architecture as the basis of service: fixed wires and fixed locations, with little financial and service flexibility.

To begin the discussion of how this technology forms the basis of service provision and institutional structure in the telecommunications industry, it is necessary to move into a more detailed discussion of the technologies of wireless communications. As soon as the discussion of technologies begins, the acronyms shoot by faster and more often than bullets at the OK Corral. Unfortunately, the author must try the reader's patience a little as we turn to the technology and physics of wireless access.

The Technological Foundations of Wireless Communications

Wireless communications are carried on waves of electromagnetic energy from one place to another, and the way in which the electromagnetic spectrum behaves largely defines the technologies that are used for wireless communications. Electromagnetic waves come in many shapes and guises, most noticeably to us as visible light. But there are a range of longer and shorter wavelengths, all of which carry energy and some of which have been brought to bear on a range of scientific and commercial concerns. From x-rays in the medical industry to radio waves in broadcasting and television, the electromagnetic spectrum is now being used by human ingenuity to carry information from one place to another.[1]

Different kinds of wireless access services are offered in different parts of the electromagnetic spectrum. AM, FM radio and broadcast television use the longest radio waves, up to about 110 MHz. Cellular services are offered in the 800 MHz band worldwide, Personal Communications Services (PCS) in the space between 1.8 GHz and 2.2 GHz. The microwave range has usually be used for point to point wireless communications. Wireless cable services are proposed for the range around 28 GHz.[2]

But there is nothing inherent about why these communications services are offered in the portion of the electromagnetic spectrum designated for their use. The designations are much more political than they are scientific, and are determined in great part by the fact that radio waves can interfere with each other if the use of different parts of the spectrum are not managed properly. For that reason, different parts of the spectrum have been assigned for different tasks. When it comes to transmitting information, bandwidth is bandwidth.

The political decisions that have determined the use of the electromagnetic spectrum were made at a number of different stages of telecommunications development and were based on honest scientific assessments of the day. But, as is always the case, old regulations are built from old science, and, even though the scientific understanding has changed, the regulatory definitions have not. Much of the public policy spawned by the technical and scientific assumptions of various periods in the past century remain with us today.

Technological innovations have made the electromagnetic spectrum seem a lot more flexible than it looked, say, 20, years ago. Whereas, before, a certain slice of the spectrum was required for a broadcast television station, the same slice can now support five or six television stations, and a few paging companies to boot.

Technical developments that allow us to harness the spectrum more efficiently open new opportunities for public policy and corporate managers. Resources that have been allocated to existing institutions, such as broadcast television, radio, and satellite can now be employed for different purposes. The institutions with access to those resources will need to transform themselves in order to reap those benefits. But any service provider will need to start with the nuts and bolts of how to build a wireless communications system.

The Components of a Wireless Communications System

When a person taps into the electromagnetic spectrum and makes a call from a wireless phone, there is a lot of different technology involved in ensuring that the call gets through. As wireless systems move from just phone calls and broadcast television to all kinds of data, voice and image transmission, the software and hardware for the systems will just become more complicated. The easiest way to understand wireless telephony without getting an degree in electrical engineering is to break a wireless telephone system down into its constituent parts and define the function of each (Balson and Macario, 1992).

For the person making the call, the starting point is the wireless phone. The wireless phone has a number of components, which can be classified into three parts: the transmitter, the receiver, and the memory and control portion of the communicator. The concept of the receiver and transmitter are simple enough for anyone who played with a crystal radio set as a child. The communicator is tuned to a specific frequency (or frequency range), picks up transmissions that occur in that range, and transmits back on that range.[3] The memory and control portions link together power sources (batteries) to the computer chips that allow for the modulation of the information to be transmitted and the demodulation of the information received.

Of course, there also has to be a device on the other end that receives and transmits the information in a way that the wireless communicator can handle. That is known as a mobile telephone switch (MTS). The MTS is connected with a transmission and reception site more commonly called a "cell." The parlance comes from a description of the territory that such as device can cover, which, on an engineer's diagram looks like a six-sided "cell." That term then became the basis for the common description, cellular telephony. As it is known today, cellular telephony is a kind of wireless access facilitated by the interoperation of a number of transmitting and receiving cells, which connect individual phones to the telephone network.

The use of cells and individual communicators presents a fundamentally different paradigm for network construction than the traditional wireline telecommunications system. An investment in wireline technology is extremely capital intensive. Not only is it necessary to connect each of the individual points on the network with a dedicated line that can stretch from 20 to 100 meters in length, but it is also necessary to put into place a trunking system that can carry all of the traffic from each of the individual subscribers. Calculating demand becomes a very precise science, and a very important one as well. Overbuilding causes financial loss that can not be recovered, and underbuilding can seriously hamper the ability of the network to provide quality service and expand to handle new traffic and subscribers.

Wireless technology is built on a scalable architecture that can be expanded and contracted much more easily to meet shifts in the demand for telecommunications services. For example, if a new subscriber base emerges in a different portion of a city or region that is serviced by wireless access, it is not necessary to rebuild the system and wire the new territory. All that is necessary is the establishment of a new cell site that can support the new subscribers.

As demand increases in highly populated urban areas with a higher demand for telecommunications services, it is possible to use a technique known as cell division (or cell splitting) to increase the capacity of the network. Instead of using only one cell to manage the subscriber base in a given territory, the area is divided up among six cells, which, in combination, increase the capacity of the network in that area by six. For a wireline network, the only option would be to rebuild the trunk lines and construct new drops from the trunk line to the homes.

As the number of customers within that specific service area increases, it is possible to divide the service territory into smaller cells, so that there is sufficient capacity to meet demand. The network can be expanded to include more MTSO's and antennas to handle the increased customer base, with smaller and smaller sites for denser and denser areas of customer demand. Scalability allows a wireless access provider to target infrastructure development to areas of greatest demand. When then demand increases, cell sites are set up, when demand decreases, cell sites can be taken down. It is expensive, but not as expensive as wires; wireline access providers do not have that kind of flexibility. Perhaps only two or three homes would choose to buy a service after the trunk line is put into place, or perhaps a competitor comes in and takes some of the customer base; the cable would have to be ripped back out of the ground, an expensive proposition, or simply be accounted for as a stranded, unprofitable asset.

As cells are split, the configuration of the network that serves the network, namely the switches that connect various cell sites together, needs to be altered to meet the demands of more subscribers. Now comes the difficult part. Creating a receiver and a transmitter is, comparatively speaking, not half as difficult as creating the link between the two items. In fact, the link between the cell site and the transmitter determines much of the engineering specifications for hardware components. The space between the transmitter and MTSO is filled with electromagnetic waves. Many different ways to ride those waves have been developed, some of which deserve explanation.

Transmission: Analog Systems

The initial mechanism for the transmission of information across the electromagnetic spectrum was analog wave forms. Analog transmission techniques are based on encoding waveforms with varying levels of intensity. The levels of intensity are decoded at the reception point and translated into speech and/or data.

The first analog systems to be made commercially available for two-way cellular communications were developed in Europe. NMT 450 was first brought into service in 1981, jointly developed and rolled out by the public telecommunications administrations of Denmark, Finland, Norway, and Sweden. (Not surprisingly, the NMT designation stands for "Nordic Mobile Telephone" and the 450 is a reference to the portion of the spectrum for which it was developed, 450 MHz). NMT 900, which operates at the higher frequency of 900 MHz, was made available in 1986. Most of the countries that have adopted NMT for their analog wireless access services are European countries.[4]

The American standard for analog transmission is AMPS, which stands for Advanced Mobile Telephone Service. Constructed by Bell Labs and tested in the 1970's, AMPS became the standard configuration for cellular systems in the United States. After the licensing of cellular carriers in the early 1980's, AMPS uses the bands reserved for cellular service in the United States, which stretch from about 820 MHz to 900 MHz. A narrowband version of AMPS, known as NAMPS, uses smaller channels for transmission and reception, which increases the capacity of traditional AMPS services. Most of the countries in North American have adopted AMPS as their standard for analog wireless services.

When cellular service was rolled out in Great Britain in 1985 (3 years after in the United States), the two providers, Vodafone and Cellnet, determined that a modified version of AMPS would be appropriate for analog wireless transmissions. The British had to modify AMPS in order to meet the European specifications of frequency allocation; although the portions of the electromagnetic spectrum assigned to wireless services in the early 1980s were similar in Europe and in the U.S., they were, nonetheless, different and required different air interfaces. It was also necessary to abide by other European technical standards that were not taken into account in the development of AMPS.

Transmission: Digital Techniques

These are the analog systems that have been the backbone of wireless services over the past 15 years. But, in the same fashion that digital transmission is fast overtaking analog transmission for wireline systems, new digital standards for wireless transmission have begun to be promoted by different interests across the globe. Instead of encoding waveforms with different levels of intensity to transmit information, the manipulation of frequencies is used to transmit the on and off, 1 and 0 of digital code. The effect is increased capacity, enhanced quality, and better connections. The three most notable techniques for digital transmission are Time Division Multiple Access (TDMA), Group Standard Mobile, or, in its French form, Groupe Speciale Mobile (GSM), and Code Division Multiple Access (CD-MA).

Before describing each of these separately, we should point out that each of these "flavors" of digital transmission have just begun their service in the industrialized world. For that reason, the equipment manufacturers that have banked their fortunes on one or another version have very strong opinions on which is better. In order to focus on the institutions behind the techniques, rather than the techniques themselves, we will try to confine ourselves to the descriptive.

TDMA is a digital transmission technique based on a relatively old innovation. During World War II, engineers learned how to isolate individual frequency channels and control those channels sufficiently to intersperse multiple transmissions simultaneously in a single channel. By dividing the channel up into time blocks (usually no more than a few milliseconds in length), and assigning each time unit to a different receiver (while transmitting in all of the time blocs), capacity could be increased. Because human speech also contains a lot of pauses and brief moments of silence, time division takes advantage of the open spaces to intersperse other information into the transmission.

George Calhoun has likened the innovation to bringing a number of people into a room and asking them to carry on simultaneous conversations with individuals on different sides of the room by choosing a specific time slot and speaking one word at a time. If people could speak as fast as computers and shoot bits to one another, the process actually might become rather fast and efficient.

To make the process commercially viable, though, requires a great deal of computer control and power. How does the communicator know which time slot is the right time slot? Or which frequency channel is the right frequency channel? By establishing a "control channel," TDMA maintains a link that is separated from the communications transmission to try to handle that portion of the equation. Obviously enough, the MTS also needs to be able to determine which conversation is which before establishing the link with the telecommunications network.

GSM is a digital transmission technique that employs a form of time division multiple access (Balson and Macario, 1992). GSM was established as a pan-European system for digital transmission, and the establishment of the GSM technique was spearheaded by the European Telecommunications Standards Institute (ETSI). By employing a non-governmental organization to codify GSM standards, the European equipment producers involved in the project signed onto a philosophy that knowledge of the system should be in the public domain (Balston, 1993). The benefits of such a commitment will include cross-country roaming, where a subscriber can travel to any portion of Europe and still use the same wireless communicator to gain access to the telecommunications network.

GSM has been largely adopted by the European PTTs and competitive service providers. Versions of GSM for other frequencies have also been developed, most noticeably the DSC-1800 version of GSM that will be used for the offering of wireless access in the 1.8 to 2.2 GHz range. Many have signed onto consortia efforts to link the North American and European TDMA systems.

The other prevalent alternative to TDMA and GSM is Code Division Multiple Access (CDMA). CDMA's answer to the problem of bringing a lot of people into a room and holding simultaneous conversations is very different from TDMA: instead of controlling everyone's conversation to a different time slot, CDMA assigns different languages to each conversation. By coding the transmissions, the communicator can pick the appropriate transmission out of the air and translate it for the user.

CDMA does not use specific channels on set frequencies, as TDMA does. Transmissions can occur over a range of frequencies, thus allowing CDMA equipment to spread through the spectrum and avoid interference in clogged parts of the assigned spectrum. The advocates of CDMA, most notably Qualcomm, fervently believe that CDMA represents a better alternative for a number of reasons.

There are other digital transmission techniques that are presently being developed, such as Space Division Multiple Access (SDMA) and variations on the Frequency Division Multiple Access (FDMA) that lie at the root of TDMA innovations. More interestingly, some claim that spread spectrum techniques will eventually replace the need for specific license assignments. The ability of transmission systems to use codes instead of frequencies allow for the use of the entire spectrum, with interference resolved through conflict avoidance technologies.

North American equipment producers have established common standards for the cellular bands and for communication on other portions of the spectrum. The initial goal of these manufacturers is to convince the cellular carriers presently using analog systems that they should make the switch to digital immediately. Some carriers have been quick to align themselves with specific manufacturers, while others have remained on the sidelines to assess the relative success of the various systems and offerings. What becomes clear, as further examination of the history and development of wireless access technologies proceeds, is that enabling technologies are as much products of science and engineering as they are products of political needs and systems. Understanding how the political realm impacts on the provision of wireless access services is the next important step in describing how wireless access can play a special role in the telecommunications development process.

The Role of Political Institutions in Managing
Telecommunications Technology

Political institutions shape technology through laws, regulations and other forms of market intervention. For the telecommunications industry, that influence has been pervasive throughout the period of PTT and public utility management. It is fair to say that the marketing function in each telecommunications company throughout the world has been subsumed under what is traditionally called external or public affairs: no service can be offered to the public without being named, classified, categorized, and regulated by the government.

That is true for access providers, content providers and equipment manufacturers. As we discussed in the previous chapter, the standards setting bodies of the telecommunications industry have certainly been hard at work establishing the accepted techniques for providing telecommunications services. There is good reason for this concern on the part of political institutions: how standards are set affect the competitiveness of individual companies and the viability of entire markets for services. That, in turn, means jobs and opportunities for constituencies to which the politicians are answerable.

For wireless access, there is one critical area of government intervention which is more critical than all others: frequency management. Governmental institutions throughout the world have taken on the responsibility of assigning frequencies, and giving those frequencies to companies and institutions is a very political decision. One might argue that it is from this fundamental issue that all regulation in wireless access is derived.

How countries resolve the allocation of frequency is by providing licenses to operate telecommunications systems, and then often require the official registration of the equipment to be used in the provision of services. Licensing has different degrees of specificity, depending on the goals of the regulators. The one element that is consistent in all countries is the identification of a specific area of the spectrum for the service to be provided and a specific geography or constituency for the provider to serve. Sometimes, it goes much further than that. The government often also specifies the size of channels and subchannels for transmission and sometimes even mandates the kind of hardware (and the manufacturer of the hardware) that has to be used to build the network.

Having defined the dimensions of the license itself, the next most evident question is: Who should receive the license, and under what conditions? For cellular systems throughout the world, most countries permitted the incumbent wireline carrier to develop, own and operate the first wireless franchise in the country. The exception to the rule was the United States, which decided to create a lottery for the licensing of cellular operators. The FCC created a duopoly in each service area, giving one license to the incumbent wireline operator and providing for a "non-wireline" license that would be given away in the lottery. (A more detailed discussion of the development of the American cellular market appears in the next chapter of this book.) All kinds of people applied for cellular licenses, and many of them had no intention of actually using the license to construct a network. The early consolidation of licenses in the most important metropolitan areas would seem to indicate that the lottery was not successful in creating a broadly diverse range of operators.

The other main option for the distribution of licenses for wireless access would be an auction where competitive bids could be tendered. The history of the auction model for the telecommunications industry is a fascinating, and has been detailed comprehensively by Thomas Hazlett of the University of California. Hazlett's research uncovers the fact that an auction strategy was first advocated by proponents of market socialism, not by the free market, anti-regulatory policy thinkers with whom the position is most closely associated today.[5] Even so, the idea was advocated by thinkers like Ithiel de Sola Pool and a variety of others as the debacle of lottery assignments became apparent in the 1980s cellular experience in the United States.

Licensing is just the tip of the iceberg when it comes to the influence that regulators exert over the market for wireless access, but it is not difficult to draw the implications and common patterns from the licensing strategy alone. Protecting a market can be achieved by limiting the number of licenses. Creating a market for a specific kind of standard requires identifying that standard as part of the license.

The interplay of technical standards, be they open or proprietary, and the licensing process becomes a critical factor in reducing the barriers to entry. There are two countervailing pressures that regulators are forced to address: In order to make a service possible, it is necessary to find some sort of technological common denominator to ensure interconnection and, where possible, universal service. At the same time, defining standards through political means is not the best way to ensure continuing technological innovation; the pressures for innovation in the market push technical advances faster than a politician's exhortation.

Corporate managers experience the same sort of tension, expressed in the paradox described by Gerd Wallenstein:

On one hand, technology-driven innovations need much interaction with targeted users before products can be standardized. On the other hand, few potential users are prepared to invest in experimental systems that may prove incompatible with network standards a few years hence. Straddling this paradox are daring entrepreneurs who engineer first applications to characteristics suitable for a few, large scale users. (Wallenstein, 1989, p. 53)

"Standardization is innovation's key to the market," he stated elsewhere in the book. That being the case, corporate managers need to determine how to integrate solutions with other market players while positioning the equipment or services they provide in a fashion that might be to their advantage.

To resolve this paradox, the industry has chosen to allow technical committees to develop standards for every piece of the telecommunications puzzle. But it is more than a matter of science. Because technologies open markets, which, in turn lead to jobs and economic prosperity, political leaders have some very unscientific interests to bring to bear to the discussion. The alphabet soup of industry and governmental organizations include many that have been established to represent the national or regional economic interests. The European Telecommunications Standards Institution (ETSI) is funded, in part, by the European Union, which is certainly interested in sustaining the competitive capacity of the PTTs and competitive telecommunications providers in the Union. The Electronics Industry Association (EIA) and Telecommunications Industry Association (TIA) in the United States may derive less of their funding from governmental sources, but their work is clearly connected to the innovations of North American equipment producers. The International Standards Organization (ISO) is perhaps the only truly international standards body.[6]

This is where the politics comes in. The typical rationalization is that if there is a national, or regional standard, and the local companies use the standard, there is an assured market for the services. That assured market will then support the expansion of the companies into other markets, where the same solutions can be imposed. In such a fashion, developing nations have often been forced to follow the telecommunications technology lead.

Wireless communications is no exception. The industry battles for the standardization of CDMA, TDMA, and GSM have been well documented in the telecommunications trade press. All three have been codified by the North American standards bodies and now that the specifications have been written down, the struggle becomes to sell transmission techniques to the service providers that will need to purchase the equipment. But that competitive ground is essentially constrained by the standard-setting function which, in turn, is based on the political arrangements codified in government regulations.

In other words, the technologies are adopted and implemented by specific institutions attempting to achieve certain goals, many of which are not consistent with the goals of development referred to throughout this discussion. It is widely agreed that the attempt by certain companies and countries to establish proprietary and protected standards restricts the diffusion of advanced technology, contravening the opportunities which would become apparent with the advent of improved access.

TECHNOLOGIES, MARKETS, AND POLITICS: DEFINING SERVICE MODELS FOR WIRELESS ACCESS

So how have the political, economic and technological factors come together in the marketplace? There are a variety of models for delivering wireless access services that exist in the telecommunications industry worldwide. As a starting point, we will take George Calhoun's description of four alternative service models for wireless access:[7]

- Wireless access as the extension of the telephone/telecommunications network;

- Wireless access as an outgrowth of cellular radio;

- Wireless access as PCN (Personal Communications Networks); and,

- Wireless as private access services.

Each of these models represents a specific combination of technology and institutional behavior. There are ownership implications, insofar as many of the systems for wireless access around the world are owned and operated by governments. In addition, there is a great deal of industry "hype" surrounding some of these models, most particularly the Personal Communications Services/Personal Communications Networks (PCS/PCN) model.

The established definitions for wireless services, most notably paging, cellular, direct broadcast satellite, and the like will be included with the individual models for the provision of services. Although the industry differentiates each of these services as having qualitative differences, it is important to remember that the same information can be carried by all different portions of the electromagnetic spectrum. In other words, any of these services, which are often offered on different portions of the electromagnetic spectrum, can be transformed to perform the same function as any of the other services. A paging company that only transmits 50 character messages today can be transmitting multimedia products tomorrow. Perhaps that kind of potential evolution of the species of the telecommunications sector is what makes this industry so fascinating right now.

Wireless Access as the Extension of the Telephone Network

Wireless services have long been seen by many as an ancillary service to traditional wireline telephony. Calhoun's first model is a reflection of that perspective, and also the reality that gave birth to the first commercial applications of wireless access in this country.

Because of the geographical dispersion of homes and residences in the United States, wireless access has been used to provide exchange services to remote areas. The main reason has been cost. The fixed and marginal cost of the wire required to link a remote area is much higher than the fixed and marginal costs of wireless services. Most monopolies with mandated service provisions have used wireless access as a mechanism for meeting the regulatory requirements in regard to rural access. The usage of wireless access for these purposes is virtually nonexistent in many of the developed countries throughout the world because the dispersion pattern of the population is not as great.

There might be other reasons as well, many of them having to do with the standards of service set by the various jurisdictions. In the United States, a number of state regulatory battles have been fought over the quality of service and tarriffing procedures for wireless access. This is because the politics of providing telephony is closely linked with the politics of politics, and a variety of options might be considered more preferable to the regulators than to the company responsible for providing the services (Calhoun, 1992).

Satellite technology has long been viewed as an appropriate technological remedy for linking remote regions into the telecommunications network (Hudson, 1990). In fact, most of the work on applying wireless access to the needs of developing countries has focused on this kind of model until very recently (Hudson, 1984). Many development programs have used as the basis of their efforts satellite technology to push levels of penetration in rural areas higher than would be possible with traditional landline technology.

The driving force behind many of these technological arrangements has been consortia of governments and international organizations, most notably INTELSAT, which helps developing countries share satellite capacity so as to expand their reach for broadcasting and telecommunications services. Examples include the Indonesian Satellite Project, funded by the United States Agency for International Development in the 1970s and a wide range of United Nations Development Program initiatives focused to meet the needs of particular countries. IMMARSAT, the international treaty organization which owns and operates a number of satellites worldwide, is focused more on extending the telecommunications network of developing countries through the provision of long- distance and international services. Those extensions of the network, though, do not add additional subscribers to the network by providing direct access to the telecommunications network.

Investments in satellite technology have become an important part of the investment strategy of a number of telecommunications companies. A wide range of industrial and commercial consortia are presently working to construct global satellite networks using a variety of hardware, transmission, and orbital techniques. The companies with the highest profile in this emerging market include Motorola, with their Iridium project, as well as the Globalstar initiative launched by Loral and Qualcomm—both of these have as their goal the expansion of the telephone network on a global scale by the end of the century. In combination, the two companies plan to spend more than $6 billion to la a projected 114 satellites (Schwartz, 1996).

Expanding the definition of dialtone to include television programming, another kind of wireless access can be included into this model. Cable television started as a hybrid form of wireless access supported by coaxial cable that brought programming to geographical areas which could not access the broadcast towers. Recently, a number of companies in the United States have begun to develop a service for providing wireless cable, which employs no coaxial cable but functions as a cable television access service (Johnson and Macomber, 1994). Direct Broadcast Satellite (DBS), although not a two-way communications medium at the present time, has boomed throughout Asia and will begin to compete directly for cable television revenues in the United States and in Latin America over the course of the next few years.

Certainly, satellite based access services for remote villages or point-to-point wireless systems that connect rural exchanges to urban centers have their place in development thinking. But they have not bridged the gap in providing telecommunications services, as the statistics presented in the first chapter clearly indicate. As Heather Hudson has pointed out, "even where countries have invested in long-distance links through leasing satellite capacity, the last mile problem remains" (Hudson, 1994; Schwartz, 1996). People are not becoming subscribers to telecommunications services, for reasons of price, technological penetration, government regulation, and corporate strategy.

For the developing and developed world, this kind of model does not go far enough. There are a number of reasons for this deficiency. First and foremost, as Calhoun explained, "what we may call the wireline derived view of wireless access tends to be cautious and pragmatic" (Calhoun, 1992). Established institutions and organizations are the focus of these efforts, and no attempt is made to create countervailing institutions that might provide access through competitive means.

Second, this model is not oriented to the possibility of replacing the existing infrastructure with wireless access services, thereby providing competitive alternatives for services in markets that can sustain competition. Integrated service providers will have to do more than simply add microwave links to the end of their networks; they will have to get into the local loops, pushing penetration levels higher.

This is certainly not to imply that such activity is a failure or has little merit. Quite the opposite is the case; thousands of rural communities now have access to services that would have been impossible to dream of decades ago. It is that focus on increasing the penetration of services which needs to be brought into our model of strategic liberalization. In linking competition and development through the strategic liberalization, we need to look to more dynamic models which foster the development of new institutions for the provision of wireless access services.

Wireless Access as an Outgrowth of Cellular Radio

In 1980, the Federal Communications Commission established rulemaking procedures for cellular operators, establishing the first North American regulatory regime for the provision of wireless access through what would become cellular telephony (FCC, 1981) A few years earlier, the Nordic countries first rolled out their systems, using a different portion of the electromagnetic spectrum but focusing on a similar market niche: mobile consumers of telecommunications services who could pay for the additional expense.

In order to bring in as many new participants as possible, the FCC mandated that one license in each territory would go to the wireline carrier (initially thought to be AT&T, but, as divestiture proceeded, the wireless assets were turned over to the Regional Bell Operating Companies) and the other would go to a non-wireline carrier determined by lottery. The number of carriers consolidated quickly as licenses were purchased in order to achieve economies of scale. By the mid-1990s, a vast majority of the "pops" served throughout the country were customers of the top 13 cellular companies.[8]

In Europe, most of the cellular systems licensed by the governments operated on a national level, and quite often there was no competition between multiple carriers until very recently. The exception to this has been Great Britain, which licensed a duopoly for its cellular systems in the early 1980s and has licensed two additional carriers for Personal Communications Networks (PCN) since that time. Few developing countries have more than one or two providers nationally, but that will likely change over the course of the next few years as countries begin to offer PCS/PCN licenses.

The goals of a cellular telephone network traditionally have been different from those of a public telephone network, a fact reflected in the business plans, government regulations and equipment produced for the purpose. Cellular telephony, in its first 20 years of development, has concentrated almost exclusively on mobile applications and high-end, business applications. Those facts form the basis of this second model, wireless access as an outgrowth of cellular radio.

Applying this model in the developed and developing world has created a very narrow business focus, and companies have been covering high fixed costs with junk bonds and a variety of leveraging schemes. The focus of the financial management of these companies has been to amortize these high fixed costs out over an extended period of time, making it difficult to begin competing with wireline phone services on cost. Even if cheaper services using the cellular model could be made available, many in the industry argue that it would be illogical to "make a $30 service available when the $70 services are selling so well" (Donaldson, Lufkin and Jenrette, 1993).

The marketing image that has been developed echoes the industry's financial need to maintain high costs by emphasizing the element of the network that produces the largest marginal cost of operation: mobility. A good example appears on the first page of the 1992 Annual Report from McCaw Cellular:

This brave new world, unfettered by a phone cord, has unleashed a tidal wave of applications that is changing the way we live, work, play and relate to one another. It means that a busy executive can travel anywhere—by car, plane, boat or any other mode—and conduct business without wasting a minute. A journalist can file a story by paperless fax from a delayed flight and still meet a deadline. An injured hiker can call for help—from a remote location. A deaf or hard of hearing individual can perform in a capacity previously not open to him. A citizen's group can report drug deals and other crimes—as they happen—and take back their neighbourhood. The possibilities are endless. The human potential—limitless. (p. 4)

The amazing growth of McCaw and other cellular companies shows that people will pay for the mobility. But the strategies have not been constructed for extensive market penetration, and, even given the rosy projections at the beginning of this chapter, there are a number of examples of how wireless access has failed to meet expectations.

The penetration of wireless access services has been retarded by the excessive reliance on the cellular service model. First, and perhaps most importantly for our purposes, the cellular model has not produced a truly competitive market that is based on price and quality differentiation. Neither real data analysis nor conjectured game theories have established that the duopolistic markets in the United States produce real competition on the basis of price. The only market in the world with more than two companies serving a given geography is the United Kingdom, which now has four separate wireless carriers. The extent of competition in that market will be the focus of our sixth chapter, but it is enough to say now that the policy and corporate communities agree that competition on the basis of price is still constrained by a number of technical and economic factors.[9]

The regulators in the United States have attempted to open the space to allow competition to evolve, and many other countries have also made room for competition within the context of the cellular/wireless access model.[10] The sticking point has been licensing and issues of market entry. If regulations and politics cannot move to create a competitive environment, then there is good reason why costs do not go down and services remain at costs that would prevent them from competing directly with traditional wireline services.

This model of wireless access as cellular telephony has not been helpful in the developing world as well. The cost structures considered profitable in the developed world have been passed down to the developing world, in large part because the developing world's cellular systems are owned and operated by the large telecommunications companies of the United States, Japan and Europe. As such, services have been largely marketed to the wealthy consumers instead of the large mass of people who do not have access to telecommunications services at all. As the *World Telecommunications Development Report* (1994) described it:

Mobile phones are often perceived to be a luxury purchase rather than an element of basic service. Certainly, if one looks at the distribution of subscribers worldwide, they are concentrated in the developed countries. The 24 industrialized nations of the OCED currently account for more than 90 percent of the installed base for mobile phones, compared with just 70 percent of the installed base of telephone main lines. Furthermore, even within these countries, the main users of mobile phones are to be found among the business community rather than the residential users. So, at face value, mobile communications have little to offer developing countries and even less to help the rural poor. (ITU, 1994, p. 37)

If the growth of cellular services is truly to reach the levels of penetration that some predict, the service will have to evolve. In the developed world, the idea that an individual (or, to match the penetration rates, 22% of the market) wants to have to deal with three phones (home, work, and mobile) seems an absurdity. To make money in the developing world, the idea that less that one tenth of 1% of a population can provide a sufficient revenue stream is even more unlikely. The cellular model, with its constrained competition and emphasis on mobility, is not a good starting point for strategic liberalization; it does not provide a path that would make wireless services a broadly based access mechanism.

Wireless Access as PCN/PCS

When the acronyms PCS and PCN were coined, many in the telecommunications policy community latched onto to the concept as a new model for the provision of wireless access services. As Calhoun put it in his 1992 book:

The third view of wireless access services is frankly revolutionary. (If this means that it is also at times rather overblown, so be it.) Radio is welcomed in its full potential as a technology of the next millennium, the communications medium of the future, the realization of science fiction dreams from *Dick Tracy* to *Star Trek*. Radio is not to be shackled to antiquated service ideas rooted in the saurian swamp of wireline telephony—on the contrary, the PCN revolutionaries rhapsodize about radio communications as a great technological imperative that will redefine our expectations about what communications services should do for us. Just as the airlines killed the railroads, so will personal, portable communications networks reduce the wireline network to a vestigial remnant, a backup systems at best, a pastime for antiquarians. (Calhoun, 1992, p. 147)

Calhoun's dramatic assertion has been echoed by many since he outlined his vision of the future of PCS and PCN. Some of it turned out to be mostly hype, other parts of that vision still resonate.

The ideal, as expressed by Calhoun, differs greatly from the business reality that has apparently settled in after the completion of the 1995 auctions for broadband Personal Communications Services licenses in the United States. As a form of wireless access, PCS has become a rather empty regulatory definition than anything else. As Craig McCaw put it in an interview in the December 5th, 1995 edition of the Wall Street Journal, PCS is "cellular at a different frequency. Adding more spectrum will drive prices down, add capacity and increase competition. It's a natural evolution from cellular."

How did this change in opinion occur? A lot of it comes down to the differences between the theory of PCS and the practice that is emerging in the United States and the United Kingdom. The theory of PCS portrayed it as the service of the future. Lightweight, easily portable headsets with digital, consumer-oriented transmission systems were developed to deliver a wide range of communications services. The construction of PCS was seen as the catalyst for single-number service, calling party pays, expanded wireless functionality through network intelligence, decreased transmission costs, and decreased consumer prices. It is also often described as the realization of "anytime, anywhere" consumer communication.

PCS did not begin as PCS, nor did it begin in the United States. In January of 1989, the Department of Trade and Industry in Great Britain published a report entitled "Phones on the Move." It identified the possibility of developing Personal Communications Networks (PCN) on spectrum unavailable to commercial users at that time. By focusing on personal rather than vehicular applications, the goal was to combine emerging spectrum and digital technology with newly developed intelligent networks (Oftel, 1993).

It came across the Atlantic more as a phrase than as an original idea. Many in the US had already been looking to new emerging technologies and the possibility of acquiring new bandwidth. On June 14, 1990, the Federal Communications Commission opened a general docket for "Personal Communications Services" issues (90-314), and began to discuss how PCS might be developed in the US. The most important questions seemed to be questions similar to those asked of cellular years before: What spectrum should it be assigned, and what technologies should it use?

This debate took place in a broader global context. The World Administrative Radio Conference (WARC) of 1992, which was sponsored by the International Telecommunication Union, allocated spectrum between 1850 and 2025 MHz and between 2110 and 2200 MHz for "future public land mobile telecommunications systems." The hope was that member nations would employ similar bandwidth throughout the world so that international standards for wireless communications would be more easily developed.

The FCC took their cue from the WARC conference and began to consider the possibility of assigning similar bandwidth to "emerging technologies," such as PCS. There was a great deal of pessimism about the potential of speedy spectrum allocation by the FCC. The cellular industry lobbied for 15 years before final definitions were given to the regulations that now govern its operation. Many believed that government would be unable to move forward on PCS.

Then politics intervened. The newly arrived Clinton administration's search for government revenues led them to the doors of the FCC. It was decided that spectrum would be auctioned so that the money could be used to reduce the deficit. Timetables were included in the 1993 Budget Reconciliation Act to insure that those revenues would be applied to the present fiscal year, and the timetables were much tighter than anyone imagined possible. Instead of waiting 15 years, the industry had less than 15 months to prepare for a spectrum auction.

But according to one observer, this auction was being driven by regulatory definitions based on technologies, "not the regulators or even the advocates of PCS" (Balston and Macario, 1992). The development of digital transmission techniques (TDMA, CDMA and GSM) and the ability of commercial equipment producers to develop hardware and software that could use higher bands of the electromagnetic spectrum has been the origin of pressure on regulators throughout the world to open up new portions of the spectrum to new competitors.

PCS services generally are offered on a different portion of the electromagnetic spectrum than traditional cellular services; cellular services were started in the 800 MHz band in the US and the 900 MHz band in Europe, while PCS/PCN services will be developed in the 1.8 to 2.2 GHz range. The higher bandwidth means that transmissions will have a shorter range, but that it will take less power to make and receive transmissions. The result is that headsets can be smaller and cheaper, and coverage can be more selective to meet the needs of the particular market niche. When first outlined, the belief was that PCS services would lower costs for wireless access in general, provoking competition within the established market for wireless access.

During the ascendancy of the idealized versions of PCS in the early 1990s, a variety of other benefits were ascribed to the PCS/PCN model. Competition was the most critical and most often emphasized element of this model. As W. Russell Neuman put it in 1992, PCS has two "hidden harvests," the first of which is the ability of a person to determine which calls to accept through intelligent network services, the second of which is the possibility of breaking up the local loop of wireline exchanges through new forms of wireless access (*Telecommunications Reports*, 1992). In 1992, Clifford A. Bean, Director of Arthur D. Little, Inc.'s Mobile Telecommunications Consulting Practice, said that "PCS is the first market-driven telecommunications offering in more than 100 years." If the second harvest which Neuman spoke of was to become a reality, wireless access carriers would position themselves in the market for local telephony and telecommunications services as a direct competitor to wireline services.

But the model proposed less than 2 years ago has evolved dramatically as licensing and commercialization of PCS services became a reality in the United States and in the United Kingdom. Many now feel that the emerging players in the market for PCS services look too suspiciously familiar to the cellular players of old, and that fact will stifle competition. In European nations such as Germany, where new systems in the 1800 MHz band have just come on line in the past few years, it is apparent that new entrants have exerted some downward pressure on prices, "but a general price reduction has still been avoided in favor of price discrimination" (Tewes, 1996).

We are back to the question posed in the discussion of the cellular service model: Why develop or charge for a $30 service if a $70 dollar service is still selling like hotcakes? The possibility in the United States that concerns the advocates of competition in the market for wireless services is that the "new" participants will develop an oligopic position that will ensure no competition on price and no direct attempt to undermine the existing predominance of the wireline network.

But even if PCS does not represent the true advent of full-fledged competition and a compete break from the model of wireless access as cellular radio, PCS does not necessarily have to turn into cellular at 1800 MHz. The United Kingdom has given us an important example of how competition on the basis of price might begin, indicating that the competitive possibilities of the PCS model have not yet been exhausted. As we noted in the first chapter, a multiplicity of companies in the marketplace does not necessarily lead to competition on the basis of price, and a limited number of companies in the marketplace does not mean than an oligopoly will necessarily arise. The four-provider marketplace in the U.K., and the potential entry of numerous others in the months to come, offers new insight into how PCS might lead to competition in the telecommunications industry, not only with the providers of cellular service, but also with the traditional wireline carriers that have dominated the market for the past century. Because the PCS/PCN model has not yet been worked out in practice, we can still take the relevant elements of the ideal and understand how wireless access services in general might emulate the ideal.

Wireless as Private Access Services

Although Calhoun did not discuss private wireless access services as a type of service model, elements of this market segment are applicable to our discussion of competitive liberalization and our attempt to identify some of the technological and service components that would be relevant to such a policy. Private dispatch services, traditionally known as Specialized Mobile Radio (SMR), are mostly used by mobile transportation companies, while paging is a service based on one-way messaging.

Both of these services have identified a specific kind of functionality that wireless access provides, constructed hardware and software offerings that exploit that functionality, and rolled out those services to a narrow niche in the market. Although there is certainly competition within these narrowly defined markets, the broader market for wireless access in general has quickly become the target for the companies that provide these service, simply because the portions of the electromagnetic spectrum that they presently use can and will be transformed into a more broadly functional kind of wireless access.

Paging was developed as a one-way specialty service for professional groups, but has evolved into a two way messaging service which does not yet include voice and extensive data capabilities. In the process, growth in the paging industry has mirrored the phenomenal growth of the cellular industry worldwide. In the United States, the number of pagers grew from about 1 million in the late 1970s to 2.2 million in 1983 and to about 10 million in 1991. During that same period, the total market for paging services grew from $1.14 billion to $2.82 billion in the United States alone (Huber, 1993). Part of that growth has been fuelled by increases in the capacity of paging networks, which has increased fivefold since 1981.

Peter Huber (1993) defines SMR is "a private dispatch service that interconnects with the public network. It has experienced rapid growth in recent years, and with the relaxation of various restrictions for interconnection, could become more of a competitor to cellular in the future." The "E" in front of the "SMR stands for "enhanced," which is a shorthand for making such networks digital and expanding their functionality so that they can provide cellular type services in the coming years.

SMR has its roots in an American definition, but SMR systems are becoming more prevalent throughout the developing and developed world. Most of the ESMR carriers in the marketplace today consist of fleet dispatch companies that provide hardware and space in the electromagnetic spectrum for cab companies, trucking companies and other mobile transportation services. Perhaps the most interesting story of the ESMR transformation from a dispatch service to a recognized competitor to wireless services is Nextel, formerly known as Fleet Call. The company, over the past two years, has attempted to project itself as a strong competitor to cellular and PCS services, even arranging for a significant investment on the part of Craig McCaw in 1996.

With the wide diversity of companies providing these kinds of products, the private service model tells us that a number of small companies that focus on niche products for specific kinds of wireless access may be able to develop a sustainable revenue stream and leverage that revenue stream to expand operations and grow into a real competitor to the larger wireless access providers. The important question then becomes, what is the appropriate niche for such a provider?

A Taxonomy of Wireless Service Models

Throughout our discussion of service models, we have sprinkled in references to different kinds of wireless access, relying on established service definitions. As has already been noted, those definitions are driven more by regulators and equipment salesmen than by marketing or engineering departments. Nevertheless, to synthesize the disparate elements, Table 3.3 presents a summary of the various service models discussed.

The irony of this taxonomy is that wireless access services are much more homogenous than the purveyors of each individual service would like to state publicly. As has been mentioned before, bandwidth is bandwidth; the services provided in one portion of the electromagnetic spectrum can be provided in a different portion of it as well. So the distinctions reflected in Table 3.2 are more by the designs of regulators and marketing departments than by real differences in the capacities of different parts of the electromagnetic spectrum.

Table 3.2:
A Taxonomy of Wireless Service Models

Type of Access	Examples	Public vs. Private	Competition Level
Extension of public network	Rural telephony via satellite or microwave	Public	Low/None
Cellular	High mobility applications	Public and private	Low to moderate
PCS/PCN	Pedestrian applications	Private	Moderate
Private wireless	Wireless LANs or paging	Private	Moderate

The market is beginning to recognize that fact as the established models and service definitions begin to bleed into one another. Cellular providers will soon be marketing "personal communications," and paging companies will likely soon be providing voice messaging. Satellites will be delivering multimedia communications, some of it one-way, some of it two-way; fleet dispatch companies will soon be offering additional data services to residential users. The list of possibilities goes on and on.

In light of this impending convergence, and because the market for wireless access is consolidating rapidly as regulatory distinctions become less and less meaningful, we need to fuse together the models we discussed to develop a generic understanding of what wireless access can do. Characterizing the future allows us to draw a clear distinction between the vision of strategic liberalization and the models that have been brought into reality by the existing government and public policy institutions.

WIRELESS ACCESS AS STRATEGIC LIBERALIZATION: LOOKING BEYOND EXISTING SERVICE MODELS

Although it would appear that there is great potential for wireless technologies to play a role in the overarching goals of telecommunications development, the preceding discussion makes it clear that existing service models make wireless access an ancillary service; either as an extension of the telecommunications network, or as a supplementary service to the fixed wireline network, wireless access still is not "on par" with wireline. As such, the potential it has as a flexible foundation for sustainable service provision has barely been tested or examined by corporate and public sector managers.

There is no established model that reflects the goals of the strategic liberalization policy outlined in the earlier chapters. There are fragments in some real world facts and conjectured examples, but there is no one real expression of how wireless access can be used to stimulate competition in the telecommunications market and simultaneously support the goals of social and economic development in the way we outlined.

Four case studies have been selected to further the search for models and ideas which are consistent with the goals of strategic liberalization. The objective in examining each of these case studies is to connect theory to reality and test some of the assumptions expressed in the first three chapters. But before launching into a specific examination of the history and development of wireless access in the United States, the United Kingdom, Russia and Brazil, it makes sense to speak in general about the kinds of challenges public and private sector managers are facing and the criterion to be used to judge which responses are consistent with the goals and objectives of strategic liberalization.

The discussion will take up that theme on two levels: the corporate response, meaning changes in the business strategy and positioning of the corporate institutions, and the public sector response, which refers to alterations in the public policy and regulatory environment that defines service provision.

Common Corporate Challenges: Decreasing Cost and Increasing Scope of Services

The ability of the industry to sustain phenomenal rates of growth is in great part due to declines in the costs for providing the services. Most of the estimates for the cost of a network are based on per-subscriber costs, and the per-subscriber costs have decreased from approximately $3,000 dollars a subscriber to at least half of that today. Wireline access is traditionally estimated to cost $1,500 per-subscriber, and it is at that cutoff point where many telecommunications companies determine whether or not a certain service will be profitable. It is this confluence of impressive growth and shrinking margins on customers that has defined the cost structure of the industry and the kinds of business practices that have been developed to support the provision of services.

Today, mobile service prices are still significantly higher than those offered by wireline service providers. When comparing peak time cellular calls to long-distance calls of identical duration, for instance, most European wireless providers today are anywhere from two to seven times more expensive than their wireline competitors (Escuita, 1996). If wireless is to replace wireline, companies will have to decrease the cost and increase the scope of services provided.

A number of factors, however, appear to suggest a long-term advantage for companies providing wireless access. The cost of constructing a wireline network is distance sensitive—the longer the wire, the more expensive the fixed cost of providing the access. The present cost structure of access services would seem to indicate that for moderately long distances, wireless access is the more economical of choices. The cost of wireless access, though, is decreasing at a faster pace. In the coming years, the costs of wireless access on a per-subscriber basis will definitely be cheaper overall, and some claim that the cost is already cheaper with the advent of the new digital technologies.

In turn, as prices for wireless access decrease, the tendency for consumers to substitute wireline phones with wireless alternatives is likely to increase. For companies with extensive wireline assets, this represents a daunting challenge: some projections even show the number of wireline telephones reaching its apex in the next few years and beginning a decline early in the next century, perhaps by as much as 2.5% (Casado, Lopez, and Sanches, 1996).

The industry faces a crossover from wireline to wireless access. That crossover has a number of common implications for public and private sector institutions: the very nature of the cost curves would indicate that wireless access can no longer be regulated as, or serviced according to the existing service models. It will not be an ancillary service if it is less expensive and represents a viable and virtually identical service offering.

But how can wireless access be made less expensive? Of all the existing service models that we have discussed, only one has been connected with the need to push down technology costs: PCS. Since PCS is heralded as the low-cost wireless access of the future, if we are looking to define a service that would push down operating expenses and thereby spur competition, we should begin with a more detailed examination of the projected costs of constructing a PCS network. The most readily available public sourcebook for these questions was written by David Reed (1992) while he was still with the Federal Communications Commission.

Reed begins by setting up a hypothesized geography of a standard layout and assumes a range of frequency blocks for providing services. At a 10% penetration rate, he estimates the total capital costs per subscriber as approximately $700 and the average annual cost of running the network as $546 per year. Although the costs would appear to be lower than the per subscriber estimate for wireline access ($1,296 in the first year of wireless access vs. about $1,500 for wireline access), the differential is not great enough to make a substantial inroad into the established revenues in developed countries. Considering the sunk cost of the wireline network, it is also not likely that the incumbent providers will take kindly to the obsolescence of their existing investment.

Reductions in administrative and marketing costs are the targets of Terrence McGarty's (1992) thoughts about how the expense of PCS services can be brought even lower. McGarty starts from the assumption that if a typical subscriber of telephone service in the United States would spend $30 a month on telephone calls, wireless access has to be able to sustain itself on that revenue stream. He assumes no economies of scale because, in his words, "wireless systems are predominantly variable in cost and they have limited fixed cost structures." The only way to reduce costs is to increase productivity and, if the revenue drops to $30 dollars per subscriber per month, the fully loaded expenses have to drop to $300 per year.

The most expensive element of all for wireless access is mobility. The complexity of the mobile switch and cell site in being able to monitor the distance of each subscriber from the base station, adjusting the strength of the transmission, handing off the call to another cell site when the person moves out of range is incredible. Being able to establish the point to point relationship for wireless access makes things much more manageable. Taking mobility out, when necessary, reduces costs.

More importantly, fixed services can be rolled out in tandem with mobile services, which provides a company the opportunity to differentiate its pricing and achieve proper rates of return for the different kinds of services. Additionally, a fixed wireless local loop application can also offer cutting edge, digital technology years ahead of the mobile systems presently in use in the developed world. In sum, this is a perfect opportunity for developing countries to leapfrog to the next generation of telecommunications technology.

What we are looking for in the case study countries, then, is a combination of cost strategies and organizational innovations appropriate to strategic liberalization. Institutional sustainability and accelerated service penetration, again, are the critical factors. The experience of cellular providers is that investment has to be amortized over an extended period of time which, in turn, negates the cost savings and potential immediate impact of new, cheaper technology. The result is a kind of business sustainability that contributes little to objectives of national development. Reductions in cost, in the end, will have to be reflected in the very structure of the institution for providing wireless access. Companies which build systems that require wide margins to sustain them in a competitive environment will not be able to make the transition to a market where margins are thinner, services are cheaper, and scale provides more of a buffer in a deeper, more vibrant market. Institutions providing services need to be constructed in a manner appropriate to this environment, prepared for the worst of competition but ready to reap the advantages of serving a broader commercial market.

Uncommon Corporate Challenges: Managing Globalization and Stakeholder Relationships

This challenge of changing cost structures would be difficult enough to adapt corporate strategies and national regulations in and of itself. But there are some complicating factors, such as the increasingly global character of the provision of telecommunications services. Companies are doing more than sustaining their own platforms, they are expanding platforms across economic and political borders. For companies wishing to play a role in the wireless access market, this challenge of increasing the scope of services is just as significant as lowering the costs of service.

Integrating those systems into global networks is certainly critical for the management of each of these companies. Common approaches to marketing, infrastructure development and service integration will provide economies of scope and scale, and common infrastructure purchase will reduce the per unit cost of equipment supply. Those kinds of globalizing factors will increasingly cause service providers to look for common access solutions and models for profitable service throughout the globe. Exchanging learning within the company becomes the critical goal. If corporations are to respond to the challenge of shifting cost curves and the decreasing cost of wireless access, there will have to be new kinds of experiments and responses to change that are integrated into an overall corporate strategy.

According to Jeffrey Wheatley (1996), there are four kinds of stakeholders critical to a company's efforts in the specific markets where an investment presence exists:

- The providers of capital, who want an adequate return;

- The customers, who want reasonable prices;

- The workers, who want reasonable pay; and,

- The society at large, requiring responsible behavior and attention to broad quality claims.

Companies wishing to define a global presence for themselves in the telecommunications market will have to address the interests of each of these stakeholder groups. With the increased degree of coordination among nongovernmental organizations throughout the world interested in issues like environmental policy, labor relations, the quality of customer service, and other areas of corporate citizenship, defining global corporate priorities will be a critical part of a company's success.

There are particular elements of wireless access that offer competitive advantage for the wireless service provider over the wireline, especially when it comes to positioning the company among a variety of stakeholders audiences—most particularly with the regulators, investors, and opinion leaders, that can make the difference between success and failure in the global marketplace. Wireless service providers should be able to manage relationships with these groups most effectively by concentrating on some of the themes inherent in the policy of strategic liberalization.

For instance, the whole concept of universal service in the developed world, or the dramatic expansion of services in the developing world, offers an immediate opportunity for wireless access providers. Quite often, governments take the first step forward by imposing community service obligations on providers so that certain social objectives are achieved. Take Rachael Schwartz's (1996) example of South Africa:

> The 1993 invitation to tender issued by the South African government advised applicants for its two GSM licenses that fulfilment of community service obligations would be a key determinant that would weigh heavily in the balance in the selection process. The Pretoria government was under pressure from the African National Congress to take actions to correct the imbalance in telephone access between the black and white communities and to increase opportunities for entrepreneurship among blacks living in the townships. Both licensees, Vodacom and Mobile Telephone Networks, have community service obligations as part of their license terms.

These sorts of service obligations are often an anathema to the competitive provider, simply because it is often thought that they erode the profitability of the company. Even so, research in the developed world has consistently shown that corporate citizenship positively correlates with customer loyalty and the willingness of consumers to pay more for services (HRN, 1996).

Such service obligations would be difficult, if not impossible, for wireline carriers to accept. But because of the flexibility of the network and the lower costs per customer, a wireless provider can use this issue to its advantage when competing against wireline carriers for the "hearts and minds" of not only customers, but also stakeholders and opinion leaders in virtually any country.

The disadvantage for wireless service providers is that the established model for wireless access is one of a premium service. Throughout the world, wireless phones are thought to be items for rich people, rather than viable options for access to the telecommunications network. Until wireless service providers begin to aggressively position themselves as potential replacement services for existing wireline networks, part of the appeal for stakeholder audiences critically interested in the social value telecommunications services can provide will be lost.

Common Challenges for Regulators:
Transforming the Regulatory Environment

Given the elements of the service model to be sought in the case study discussions, what might be the challenges for the regulators wishing to promote sustainable competition and development through wireless access?

As has been mentioned before, wireless carriers throughout the world face three kinds of regulation: frequency licensing, price and technology. Perhaps the most significant of these is the first; the number of licenses that a country chooses to distribute determines the competitive structure of the provision of services. Until the conclusion of the PCS auctions in the US in 1997, no country in the world has more than four wireless access providers operating in a single geographical area because of licensing restrictions.

Price regulation is significant inasmuch as certain localities in the United States and many countries have set the rates for calls by mandates, which, in turn, defines how the system can be operated. Prices are often set with an upper limit and the high fixed and operating costs of running a network using the present cellular models forces carriers to keep prices high while the costs of network construction are amortized over an extended period of time. Closely related to price regulation is the problem of interconnection; the costs of connecting a call to a landline network determines the price that a wireless access provider can offer to customers, and the mandating of interconnection fees is often considered to be an important part of ensuring that there is no cost advantage for wireline carriers who might attempt to cross subsidize their operations to offer lower prices.

Technological mandates also provide handcuffs by dictating the basis of network construction. Many countries have dictated that licensees will use a specific technology, such as GSM in Europe. Restrictions in this area are much looser in the United States, where carriers have been permitted to use a range of hardware and software solutions. Wireless carriers that face a tight regulatory structure in all three of these areas face a market that has been totally defined for them.

To begin by way of a comparison, the table 3.4 outlines the regulatory character presently in place in a sample of countries. What is clearly apparent is that countries with earlier initiatives in privatization are, by and large, much more open in their regulatory stance towards wireless access.

The bottleneck for wireless access is, as always, the problem of market entry. There is a broad commonality in the regulation of wireless access worldwide, with New Zealand offering perhaps the most open system for new entrants into the market, at least on paper. But in every other country around the world, licenses are the key to new market entry, and granting licenses has been a political, not economic, decision. If there is no threat of entry, competitive pressures are greatly diminished.

If competition is to be the goal, then regulators must primarily solve the problem of market entry. In one respect, this means the licensing of new competitors. Identifying technological opportunities to use different portions of the spectrum are an important first step, and the development of PCS is due to the technological innovations that permit the provision of wireless access at 1.8 to 2.2 GHz.

Table 3.4: Comparison of Wireless Access Regulation

Country	Controls on wireless entry	Controls on wireless operation	Wireline privatization initiatives
Canada	Closed duopoly	Federal authority exists, but rates not regulated	Wireline carrier is mixed public/private ownership
France	Closed entry — duopoly with special licenses	Technical requirements, resale of excess capacity	France Telecom to become a corporation in 1997; 1998 EU market liberalization deadline
Germany	Five licenses	Spectrum user fees charged	Privatization initiated in 1996
Mexico	Closed duopoly; Telmex in all regions, second carrier private	Telmex rates regulated through SCT	Privatized in 1991
New Zealand	Open in principle, with four licenses and two operators	No controls	Privatization in 1987
UK	Closed entry, duopoly for cellular and for PCN. Multiple licenses issued for wireless local loop, data	No controls	Privatization in 1984
US	Closed duopoly for cellular, multiple licenses for PCS	Limited federal regulation under the 1996 Act. States may regulate services	Industry has always been private

But giving out those licenses in an economical fashion is equally as important, which is why an examination of auctions and new licensing mechanisms in the case study countries will be an important focus of the discussion. Auctioning licenses gives regulators an opportunity to achieve a solid valuation for the use of the electromagnetic spectrum, while simultaneously insuring that the government receives finds for the allocation of spectrum.

In another respect, reducing the barrier to market entry means the creative redistribution of wireless service licenses that presently exist. For developed nations that have granted broad swaths of bandwidth to broadcast television stations, it will be important to allow new opportunities to provide cross over services. For developing countries with a more limited market for such services, it will be critical to bring together resources so that these kinds of advantage can be exploited.

Barriers to entry are also driven by costs, and costs, as we have seen, are also closely tied to regulation. In the cases where regulators proscribe certain technological solutions for network construction, the mandate is often tied to the thought that equipment manufactures, having standardized systems to produce, will be more likely to achieve economies of scale. Costs of network construction can thereby be reduced.

Nevertheless, regulators quite often codify standards in a fashion that also codify the cost of providing those services. Mobility, for example, is an essential part of the GSM standard, and the possibility of a company providing a fixed network service using GSM would be possible, but there may be lower cost solutions that are developed for fixed wireless access services. In addition, codifying standards creates vested interests in the equipment manufacturing business, and the introduction of new technology can be stifled.

For those reasons, lowering the barriers to entry also requires a more flexible approach to technological standards worldwide. The codification of alternative standards is clearly a useful exercise, but mandating certain standards may serve to retard the introduction of new technology by keeping costs higher.

Unique Challenges for Corporations and Regulators: An Overview of the Case Study Countries

Having outlined the specific attributes of the wireless "species" we will be hunting for in the global telecommunications jungle, is it perhaps worth describing why a case study approach has been chosen and to talk a bit about why certain countries have been selected over others for the purposes of this study.

The provision of telecommunications services has largely been a national affair. Although there have been significant local and international efforts to provide certain kinds of communications links, the predominant concentration of investment and regulation have been on national levels. This is less true in the United States than in other countries, but it nevertheless indicates that the most appropriate starting point for the comparative analysis of telecommunications structures is likely to be at the national level.

In doing so, there is a clear analytical link to the discussion in the previous chapters. In this period of nation-states, much of the resources that can be directed toward the goals of economic and political development are centered at the national level. Additionally, the most significant comparative studies that have been published have used geography and regional commonality as the basis for comparison and contrasts. A national approach allows us to make a clean connection between the dominant writings in the literature of development and the case studies we articulate.

Most of comparative political theory has focused on grouping together countries with common characteristics. The most notable, and certainly the seminal example of the application of comparative theory to an understanding of political relations, is based on such an approach: *The Civic Culture*, written by Gabriel Almond and Sydney Verba in 1963, uses as its take-off point social science research from a range of democratic countries. Their goal in writing the book was to assess the various cultural factors that affect the particular expression of democracy in each of the countries examined.

But which case studies do you choose in launching such a comprehensive study? The similarities and dissimilarities of culture lend themselves to groupings and aggregations that may not be appropriate for other analytic purposes. The goal is to show the full range of contexts when strategic liberalization can be of use. The sample, therefore, needs to be broad and even, comprised of both developing and developed world contexts.

To achieve that balance, this book looks at comparisons and contrasts and chooses countries so that there are prima facie elements of similarity and difference. The case studies presented include two developed countries, one developing county and one country undergoing a transition to a post-communist economy and polity. This balance allows us to bridge the major socioeconomic gaps as they have been defined by the academic literature since the advent of the cold war: We have examples from the nominally defined first, second, and third worlds.

The two first world countries share an important commonality when it comes to the provision of telecommunications services: Both have been at the forefront of privatization and deregulation throughout the world. The United States, with its emphasis on managerial capitalism and private ownership, has always relied on a private company to provide telecommunications services, and took a major leap in making the market for such services competitive over the past decade. So, too, has Great Britain worked to privatize its formerly state-owned telecommunications provider and inject further competition into the marketplace for services.

There are two significant differences between the development of the markets of these countries that makes them optimal choices for comparative analysis. Most evidently, there is a great difference in scale when it comes to the two markets. The size of the United States, as compared to Great Britain, creates very different conditions for the development of a competitive marketplace for telecommunications services.

Additionally, there is a cultural difference between the countries that is expressed in the form of corporate organization. Alfred Chandler has identified this as the difference between the managerial capitalism of America and the personal capitalism of Great Britain. Although we will not use Chandler's distinctions as the basis for describing structural differences between the providers in the United States and the United Kingdom, it is important to recognize that differences in business practices and cultural norms do impact on forms of industrial organization.

The choice of Russia allows us to look specifically at a country that is making the transition from a totalitarian, communist society to a new social and political order that has yet to be defined. Moving from the total state with complete ownership of all of the means of production to an open, competitive environment for goods and services is a monstrous transition. Understanding this piece of the puzzle will provide us significant insight into how public policy and corporate management can take advantage of the opportunities afforded by such a transition.

Brazil and Russia are similar in terms of scale, but their political and economic histories are highly divergent. Brazil's consistent struggle with the establishment of democratic forms of governance and its relentlessly bureaucratic politics among the social and economic elites provides a dramatic contrast to the situation in Russia. They start from a similar point in terms of service penetration and other measures of telecommunications performance, but the means of continuing improvement in network expansion and improvements in the quality of services are very different indeed.

Details on both the development of wireline and wireless telecommunications services are provided in each of the chapters that follow. As the chapters progress, more comparative discussion is offered to draw out themes and trends that stretch across the developing and developed world.

STRATEGIC LIBERALIZATION AND INSTITUTIONAL CHANGE

The service model appropriate to strategic liberalization would be a new species in the jungle of the telecommunications sector, but one that could transform the rest of the jungle by example. As a strong competitor, it would prey on some of the weaker, less agile and nimble companies unable to keep up with changes in the cost curves. It would intermingle with other types of species, combining, providing the foundation for the jungle's equivalent of strategic alliances and new kinds of evolution. To borrow an expression from evolutionary biology, it could play a critical role in the period of change and upheaval that is inevitably coming in the telecommunications sector, punctuating the equilibrium that has marked telecommunications development to date.

The institutional relationships that have defined telecommunications development in each of these countries are based on a common perception: Centralized forms of development are the best mechanisms for telecommunications. For institutional change to take place, there needs to be a challenge. New technologies are part of the challenge, but until institutions arise that can harness the new technologies, the existing structural relationships that define telecommunications development will remain unaltered.

The model being sought in the case study examinations is, in some ways, the inverse of the centralized approach. Working cell by cell and access point by access point, a telecommunications infrastructure can be built from the ground up to provide a competitive challenge against the existing economic and political institutions. By identifying a range of appropriate corporate and public policies, the discussion has sketched out a way to make these newly arising institutions sustainable.

In doing so, strategic liberalization is potentially a revolutionary attempt to force institutional change. By fostering the growth of wireless access and allowing new institutions to take form around the more flexible, less costly approach to telecommunications access, it is possible to induce change and move forward more quickly on the path to telecommunications development.

What becomes clear from the discussion in this chapter is that the policy of strategic liberalization will be different in different contexts, and will have to draw on the specific historical and technological realities of each country. Certainly, applying some of the elements of strategic liberalization in the chaotic markets of Russia will lead to dramatically different results than in the highly centralized and regimented administrative bureaucracies of Brazil.

Nevertheless, there is enough conceptual coherence to draw certain principles from each of these situations including the possibilities of identifying technological innovation and rewarding those innovations; the need to overcome the barriers of burdensome regulation and administration; and the possibilities for price competition and the use of appropriate technology. With those thoughts in mind, the detailed discussion of each country begins.

ENDNOTES

[1] The science of wireless access is presently under scrutiny after reports that cellular phones may cause cancer. Extended exposure to certain parts of the electromagnetic spectrum can certainly be harmful, such as x-rays and other forms of radiation. The ongoing debate on Electromagnetic Forces (EMF) has impacted on the market for wireless access and some consumers wait for more conclusive scientific evidence.

[2] Different kinds of waves have different modulation patterns and modulation frequencies. The "Hz" designation is a measurement of the number of modulations per second, with MHz an abbreviation for "megahertz" and GHz an abbreviation for "gigahertz." The three services mentioned as examples are ordered from the longest waves to the shortest waves.

3   This is a vast oversimplification, especially considering new innovations in what is called spread-spectrum technology. George Gilder has claimed that the advent of spread spectrum technology will completely remove the need for the assignment of frequencies of the electromagnetic spectrum.

4   The countries that have adopted NMT include: Andorra, Byloerussia, Croatia, Estonia, France, Indonesia, Luxembourg, the Netherlands, Poland, Saudi Arabia, Sweden, Tunisia, Austria, China, Cyprus, Faroes, Hungary, Latvia, Malaysia, Norway, Romania, Slovenia, Switzerland, Turkey, Belgium, Czechoslovakia, Denmark, Finland, Ireland, Lithuania, Morocco, Oman, Russia, Spain, Thailand, and Uzbekistan.

5   Hazlett attributed the idea of applying auctions to the allocation of spectrum to Leo Herzel, not Ronald Coase as some have suggested. The association with market socialism comes from Coase, as Hazlett pointed out in his article by drawing on some of Coase's commentary on the subject:

> It is sometimes said that I introduced the idea of using prices to allocate the spectrum. But this is untrue. The first time this was proposed, at any rate in print, was by a student author, Leo Herzel, in an article in the University of Chicago Law Review in 1951. When I first read this article I thought, and it was quite natural to think this, that Leo Herzel had been influenced by Aaron Director and Milton Friedman. But this is also untrue. While he was an undergraduate, Herzel had become very interested in the debate over whether a rational, efficient system for allocating resources would be possible under socialism. As a result, he read Abba Lerner's *The Economics of Control* soon after it was published in 1944. This debate, particularly Lerner's detailed proposal for market socialism in *The Economics of Control* was the inspiration behind his views.

The point emphasized in the above quote is clear: the philosophical underpinnings of auction policy have more to do with market control than market freedom. In that regard, auctioning could be described as rent seeking behavior by governments who assert access over the resource of the electromagnetic spectrum. See Thomas W. Hazlett, "Assigning Property Rights to Radio Spectrum Users: Why Did FCC License Auctions Take 67 Years?" Telecommunications Policy Research Conference, September 1995.

6   Just because these bodies establish standard technical solutions to telecommunications problems does not mean that they will become widely accepted. Quite the contrary. As international bodies with no power to enforce their decisions, quite often their solutions are deemed by the marketplace to be incorrect. The most evident example is the immense success of TCP/IP as the internetworking standard of the internet, which was developed separately from the ISO's mechanism for data transmission and networking.

7   This taxonomy is taken from George Calhoun, *Wireless Access and the Local Telephone Network*, section 5.2 (starting on page 135). The fourth alternative is suggested more as an ownership distinction than as a service model in Calhoun's work.

[8]   The cellular industry commonly uses "pops" to describe the number of individuals that can be served by a particular cellular license. Short for "population," the term often assumes that a particular license can and does in fact serve every single person in a region, which certainly does not correspond to the business reality of building a market for an emerging service.

[9]   See, in particular, the proceedings of Docket 91-34 of the Federal Communications Commission on the bundling of customer premises equipment (CPE) and wireless services. The Department of Justice's criterion for a workably competitive market is based on the Herfindahl-Hirshman Index's assessment of market concentration.

[10]   The federal bent to cellular telephone regulation is evident both in legislation and regulatory decisions on the matter. As Peter Huber et al., write in *The Geodesic Network II*:

> The [Federal Communications Commission's] procompetitive policies have made it unnecessary to regulate [many] aspects of radio services. The rates, revenues, and profits of radio service providers are subject to no federal regulation, and most states do not regulate cellular or paging providers at all. Most states that do regulate such services do so only to a very limited extent, requiring such things as informational tariffs, and typically imposing no price regulation at all at the retail level. The competitive policies licensed upstream, in allocating licenses and overseeing equal interconnection with the landline network, make additional regulation unnecessary.

See, in particular, *In re Revision of the Uniform System of Accounts and Financial Reporting Requirements for Class A & Class B Telephone Companies*, 60 Rad. Reg. 2d (P&) 1111 (FCC 1986).

Through These Portals:
Strategic Liberalization in the
U.S. Context

The future of telecommunications policy in the United States has become wrapped up in one piece of legislation: the 1996 Telecommunications Act. The Act has been hailed by politicians as the final breakthrough for competition in the U.S. telecommunications industry, and is the centripetal of reference for the public communications of major telecommunications companies as they announce their plans for mergers, takeovers, and the introduction of new products and services.

The passage of the Act represents, in some ways, a significant change in the way telecommunications networks are talked about in the U.S. Visions of a "network of networks" and a unified "national information infrastructure" driven by public sector investment and regulation has largely given way to discussions of individual networks and corporate strategies. That is not as it should be—there is still a need for an overarching vision of how the U.S. telecommunications infrastructure can evolve and support the continuing social and economic transformation of the country, if only to allow for the coordination of technologies and infrastructure development programs undertaken in the private sector. In that regard, the dialogue on the future of a national information infrastructure (NII) for the U.S. will be difficult to resolve, even after all the provisions of the Act are brought before the Courts for interpretation and through the U.S. Federal Communications Commission for implementation.

A portion of that discussion will involve the role of wireless technologies and how the increasing use of cellular, PCS, satellite, and other wireless access services in the U.S. will contribute to the evolving competitive marketplace. At the moment, there is little discussion on what that role might be and how wireless access can serve as a driver of industry transformation during the upcoming years of industry change. The thinking in the regulatory mainstream mostly reflects the language used in a 1995 report by the Office of Technology Assessment of the U.S. Congress, entitled "Wireless Technologies and the National Information Infrastructure." Broadly speaking, the vision of the paper can be summed up in these two paragraphs which appear in the first chapter:

> Wireless technologies can extend the NII in two important ways. First, they allow users to tap into communication and information networks as they move about. Mobility is a key driver for wireless. Second, as noted earlier, wireless technologies can extend NII services to places where wire is too costly or difficult to install. This may prove to be especially important as links need upgrading. In this role, wireless systems will help ensure that future universal service goals are met.
>
> Wireless technologies and systems will also compete in the delivery of NII-related services, both among themselves and against wire-based services. Competition is a key principle underlying the NII, and different wireless services have advantages that will allow them to compete effectively in a number of markets. For example, Broadcast, DBS, and Multichannel Multipoint Distribution Service (MMDS), already compete with cable television systems (and each other) across the country, and competition is expected to increase as companies convert to digital and new competitors enter the market for video services. Wireless technologies are also expected to make a substantial impact in the market for voice and data communications, especially where mobility is desired. A good deal of spectrum has recently been allocated for wireless voice and data services and companies have been working on systems for a number of years. Many analysts believe that wireless could become the voice communications technology of choice—eventually becoming a substitute for existing telephone service—because it offers the added advantage of mobility. Over the next five to 10 years, wireless technologies will emerge as significant competitors in most communication, information and entertainment markets. (Office of Technology Assessment [OTA], 1995, p. 32)

The conclusions of the OTA are, to put it mildly, unimaginative. At the core, the authors start from the existing cellular and PCS model in assuming that mobility as it is presently offered through cellular services is the critical factor for the replacement of wireline technologies by wireless technologies. Either as an extension of the existing infrastructure, or as a direct competitor to wireline services, consumers will not begin to make the transition until the prices for wireless access are below those for wireline access. Unlimited mobility will be priced as a premium by companies for as long as possible to keep margins as thick as possible. For that reason, the model has to shift if wireless is to take its place as a complete contributor to the NII.

On a more fundamental level, the report misses the opportunity to embrace what could become a revolution in wireless telecommunications services: the complete remaking of the architecture for the provision of services. It assumes severe limitations on bandwidth capability (and availability), and relegates wireless access to a lesser role, competing only for specific kinds of services rather than as a broad backbone for mass access to the NII. Wireless access can become the basis for facilities-based competition, serving as an access technology to sustain corporate institutions wishing to serve the telecommunications market. That more compelling vision should be a central part of the ongoing discussion of the future of the NII in the U.S., shifting the terms of the debate to include the vast array of potential resources that this jungle of telecommunications products and services may be able to offer.

This chapter, like the other case study chapters that follow, is divided into three parts. The first provides a brief overview of the development of the telecommunications sector in the United States, focusing specifically on the regulatory and structural characteristics critical to our comparative analysis of the various telecommunications infrastructures. An examination of the history of wireless access in the U.S. follows, leading to the final portion of the chapter, which returns to the proscriptions of strategic liberalization and the possibilities for the sustainable development of a truly national information infrastructure.

THE U.S. MARKET
FOR TELECOMMUNICATIONS SERVICES

There are two tensions that will shape the telecommunications sector in the United States long after the passage of the 1996 Telecommunications Act: the tension between universal service and competition, and the bifurcation of regulatory policy and corporate operations between the federal government and the individual state governments. These two tensions are at the core of the U.S. experience of corporate and political governance, so it is not surprising that we meet them again within the context of a discussion about the country's telecommunications development. In fact, it might not be overstating the point to agree with one commentator who characterized the "regulatory chaos in [American] telecommunications" as "essentially a manifestation of a deep-seated cultural pattern."

> Americans have always had a love hate relationship with 'centers.' In 1832, de Tocqueville observed that 'people wish to keep the Union, but to keep it reduced to a shadow: they would like to have it strong for some purposes and weak for the rest—strong in war and almost nonexistent in peace—forgetting that such alterings of strength and weakness are impossible.' They have repeatedly exhibited great discomfort with the emergence of any authority structure that could impose order in a centralized manner.

> Because of this dispersion of authority, the USA has always had difficulty in creating the infrastructure of the day—canals, railroads, telegraph, electricity, highways and telecommunications networks. The development of infrastructure networks requires placement of interlinked pieces of technological hardware over geographical space. The individual pieces of hard-

ware have never been a problem in the USA. The problem has been of an organizational nature. (Sawhney, 1993, p. 506)

This "decentralized scene of considerable confusion" (Lee and Cole, from Dutton, Blumer and Kraemer, 1987) has been the hallmark of American telecommunications development. The confusion over goals has been a substantial part of it: universal service or increased competition? On the other hand, there is the confusion that comes from a federal system of government: local or national jurisdiction? The persistence of these conflicts has largely determined the development of the telecommunications networks of the United States, and is reflected in the developments in the sector over the past twenty years.

Needless to say, any strategy for telecommunications development needs to suit the national character, which is perhaps the most compelling reason to combine the best elements of comparative politics with an analysis of telecommunications development. In the American case, the goal is to define a strategy that is decentralized so as to ensure the participation of local interests even with the direction of national interests. Considering the size of the market, and the increasing complexity of the marketplace, this is quite a daunting task.

In many ways, the framework laid out in the 1934 Communications Act was enough of a marriage of the two to sustain itself for an extended period of time. Under the guise of economies of scale and economic necessity, the establishment of AT&T as a private monopoly represents an unique consensus in a culture that finds it difficult to implement any public sector system for any kind of service. With AT&T established in place after the period of access competition in the United States, the telecommunications development of the country was driven by central investment dynamics similar in many ways to those that in other countries with state-run telecommunications companies.[1] But it was different in one fundamental fashion: The goal was to return value to shareholders, not financial gain to government pockets.

Perhaps it was that one difference that allowed the United States to maintain a sizable lead in telecommunications for such a long period of time. AT&T, through its subsidiaries, serviced the needs of local consumers (through the 22 Operating Companies), long-distance services (through its long lines divisions), and produced the world's most advanced telecommunications equipment (most of which was pioneered at Bell Labs). The Bell System had a culture all its own, some of which can be seen in the buildings now used by the regional bell operating companies that represent the local legacy of AT&T. In New York, for example, right in the middle of the financial district, one of the old Bell buildings still stands, with its ceiling tile frescoes of telephones that echo the themes of Michelangelo's Sistine Chapel. Above the door leading to the 29th floor conference room is written: "Through These Portals Pass the Best Telephone People in the World."

That all seems far behind us now. There is no longer one institution with a monopoly on telecommunications services, much less the best telephone people in the world. New portals are being developed, all resting on the uneasy foundation of a culture split between universal service and competition, between local and national authority.

The next two sections focus separately on the regulatory and marketplace changes since divestiture, emphasizing the role that these fissures play in the reactions and strategic planning of government and corporate managers. We then conclude this section with an outline for how strategic liberalization can be targeted to address the needs of telecommunications development in the American context.

From the Modified Final Judgment to the Present Day: The Limited Regulatory Compact

Before the first break-up of AT&T in 1984, the history of telecommunications regulation was, to be frank, highly technical and largely uneventful:

> For most of the forty years following the passage of the Communications Act in 1934, the most visible and significant questions of communications policy were largely questions of broadcasting policy. Questions of telecommunications (that is, telephone or telegraph) policy, when they emerged, were generally resolved through negotiations with American Telephone & Telegraph or Western Union.

> While commission policy in all of these matters was appealed to the courts (and in more than a few instances, to the U.S. Supreme Court) and was occasionally subject to review and revision by Congress, the development of communications policy after 1934 was generally left to the regulators and the industry. Arcane communications issues were only dimly (if at all) perceived by the public, and there was little political gain to be had from involvement with them, with one exception: as the power and importance of the electronic media in the political realm increased, members of Congress became increasingly attentive to agency decisions affecting broadcasters in their districts. (Symons, from Newberg, 1989, p. 275)

Antitrust issues had always been at the core of AT&T's relationship to the political community, and there certainly was some political interest in the telecommunications industry. But the constant parade of politicians claiming knowledge of and interest in telecommunications policy and investment was lacking as compared to the last few years. Basically put, telecommunications policy had yet to move from closed-door discussions in smoke-filled rooms to the veritable street brawls of recent telecommunications legislation.

The Modified Final Judgment (MFJ) changed all that. In 1956, one of those quiet backroom deals had been made between AT&T and the Justice Department, resulting in a consent decree that compelled AT&T to ensure the cross-subsidization of services between portions of the company did not affect local telephone rates. But this agreement became increasingly less tenable as new technologies

and new kinds of competition slowly undermined the political and economic viability of the system. The history of litigation regarding the activities of AT&T
from the period between the Kingsbury commitment and the divestiture of the
company is well documented and does not need to be reviewed again here. It suffices to say that the pendulum that had swung heavily in favor of universal service
in the beginning of the century had begun to swing back to open the opportunities
for increased competition.

The resounding statement of this shift came from halls of the District of Columbia's Circuit Court of Appeals, under the direction of Judge Harold Greene. He
gave his blessing to a modification of the 1956 AT&T consent decree, and divided
the company into two parts. One part of the decision reads as follows:

> The proposed decree would provide for significant structural changes in
> AT&T. In essence, it would remove from the Bell System the functions of
> supplying local telephone service by requiring AT&T to divest itself of the
> portions of its twenty-two Operating Companies which perform that func-
> tion.
>
> The geographic area for which these Operating Companies would provide
> local telephone service is defined in the proposed decree by a new unit, the
> "exchange area." According to the Justice Department, an exchange area
> "will be large enough to comprehend contiguous areas having common so-
> cial and economic characteristics but not so large as to defeat the intent of
> the decree to separate the provision of intercity services from the provision
> of local exchange service." Court approval would be required for the inclu-
> sion in an exchange area of more than one standard metropolitan area or
> the territory of more than one State.
>
> The Operating Companies would provide telephone service from one point
> in an exchange area to other points in the same exchange area—"exchange
> telecommunications"—and they would originate and terminate calls from
> one exchange area to another exchange area—"exchange access." The in-
> terexchange portion of calls from one exchange area to another exchange
> area [would be provided by companies] such as MCI and Southern Pacific
> Co. (*U.S. vs. American Telephone and Telegraph Company; Western Elec-
> tric Company, Inc.; and Bell Telephone Laboratories, Inc. F. Supp. 131*)

Most of the reasoning that appears in the document is constructed under the
guise of economic principles. This is always the case in antitrust litigation, which
has always at least tried to start from the basis of economic principles arising from
the imperative to sustain a market of many small-to-medium-sized competitors,
rather than a market dominated by few.[2]

It is clear that the MFJ was as much a political decision as an economic one.
Because this "modification of final judgment" was proposed by AT&T and agreed
to by both Judge Greene and the Justice Department, it should not be surprising
that the decision is imbued with the political tensions that define America's tele-

communications development. But the decision reacts in a fundamental way to the tension between local and national authority, and between universal service and competition. The problem is that it leaves the issues unresolved, suspending the conflict between the two sets of opposing principles. The MFJ therefore represented a limited compact, one that was bound to be undermined as market participants determined how to position themselves in the emerging environment.

For example, the basis of the justification for an establishment of a monopoly for service provision in the United States was the "universal service" mantra coined and communicated by AT&T as its corporate strategy. The need to sustain that vision for the country's telecommunications infrastructure is reflected in the burden placed on the Regional Bell Operating Companies (RBOCs) that were created through this decision. They were to be the vehicles for universal service by functioning as exchange operators.

History has taught us that this is an illusion, and the 1996 Telecommunications Act is a recognition that the regulations no longer fit the times. In great part, this was a necessary illusion, but it appears that many of those who participated in the decision recognized it as such. Even Judge Greene wrote in the final decision that AT&T might choose to bypass the Bell Operating Companies that had just been created to offer local services directly to customers. In Peter Huber's (1992) words, "Judge Greene dealt with the problem by wishing it away," thereby establishing a space for the idea of universal service and local autonomy to sustain itself.

The introduction of competition was also construed in such a fashion as to freeze the ongoing tension between local and national authority over telecommunications regulations. From the economic point of view expressed in the MFJ, the liberalization of the long distance market represented the best possible option for the introduction of competition in the telecommunications sector, given the technological and organizational state of the market participants. But also it was the only kind of competition palatable to both local and national authority of the time.

On the local level, competition in long-distance did not undermine in any substantive way the significant power of state and local governments. It had always been the state's prerogative to regulate rates and, in turn, dictate investment policy to the individual operating companies. By organizing intraexchange services into local jurisdictions, the MFJ ensured that telecommunications boundaries mapped state boundaries, because there are no LATAs that are part of more than one state. The state Public Utility Commissions and Public Service Commissions would (and do) still regulate the RBOCs on a state by state basis, and the MFJ was implemented in large part because it avoided potential local roadblocks.

On the national level, the goal of competition could be met through a national infrastructure: long-distance networks. The decision also provided for a complete and unified information infrastructure, an issue critical for U.S. national security. The FCC quickly took on the role of managing the interface between the local exchange and the long-distance companies, thereby defining the outer limits of what was nominally competitive and nominally granted to universal service.

So it might be said that the philosophy for telephone systems has been, in the American case, as follows: Local shall be governed by universal service, and national services shall be competitive. On occasion, Congress and the President tried to remake that vision, but little significant legislation passed through Congress in the decade after the MFJ was put into place. Congress did succeed in regulating, then deregulating, then reregulating the cable industry during the late 1980s and early 1990s. The flip-flopping of regulations did little to improve customer service and prices and served to undermine some of the confidence of the investor communities in the big cable companies. Other than that, the talk of telecommunications reform had been like the passing of the cherry blossoms in Washington: blooming majestically every year only to die someplace between the Congress and the White House. Even with a Democratic Congress, a Democratic President, and a Democratic Vice-President with a substantive knowledge of telecommunications issues, the telecommunications reform package of 1994 died, and this in a year when everyone was talking about the "telecommunications revolution."

In 1996, though, the powers came into alignment. Surprisingly, a Democratic President and a Republican-controlled Congress were able to accomplish what had not been accomplished since 1934: New telecommunications legislation was passed and signed into law. Taken at face value, the new law allows for telecommunications companies to make significant inroads into each other's territories, permitting former monopolies, like the Regional Bell Operating Companies, enter the market for long-distance services and local entertainment or cable services after certain regulatory hurdles are cleared. But less significant than the actual legislation itself is the corporate responses to the opportunities that the legislation makes available.

Preparing for the Next Battle:
The Corporate Players in the U.S. Telecommunications Sector

Even with the passage of the Act, the corporate strategies that have been pronounced by American telecommunications and information companies over the last 5 years all boil down to one basic statement: Each company wants to own its own platform and invest in the value of that platform.

There is good reason for this to be the strategic mantra of the telecommunication industry in the United States today. History tells us that telecommunications companies will live or die in their ability to gain access to and directly serve customers. That is impossible without control over and ownership of the technology, the information content, and management structure to make the "platform" valuable enough to consumers. The fact that all of the players in this industry have come to this strategic conclusion can only mean one thing: They have seen the future, and it is in the competition between various, separately managed but physically interconnected, service platforms. The MFJ, in many ways, has become the illusion that was required to prepare the way for the battle that will now begin to unfold with the passage of the 1996 legislation.

The players, though, have started from different points and with different kinds of resources and investments already made. Before moving on to address how the market for wireless services has developed and will impact the progress of competition within the telecommunications sector, it is worth discussing the various positions of the wireline service providers as they struggle to define and enhance their own service platforms.

Long-Distance Carriers

The MFJ certainly has had its resiliency, in great part because the decision also reflected the needs of AT&T's corporate culture at the time. Not surprisingly, the contrast between local and national interests was even a significant part of AT&T's character as well. The company's 22 operating companies often pushed for independence vis-à-vis the center, and certain operating companies were predominantly overrepresented among senior management in corporate headquarters. By letting go of the operating companies, AT&T was able to shed the federative structure imposed on the company through its relationships to the operating companies. The divestiture of the equipment producing arm of AT&T, which was also considered as an option during the proceedings, would still have left the company in the difficult position of integrating local interests with national ones, a tenuous position at best.

But AT&T decided to turn around and do just that in 1996. AT&T split itself into three separate companies, one for its communications services, a second for its equipment production division, and a third for its computer business. The new AT&T communications services company, though, will continue forward in the strategy of creating an integrated service platform that reaches customers at the local level, unencumbered by the need to position itself as a company that sells equipment to those with whom it is about to compete.

After the MFJ, there was much talk of the emergence of competition in the long-distance market in the United States. The growth of other network providers, such as MCI and Sprint, has changed the telecommunications sector and offered a new range of possible investment. Projections were that AT&T would lose a substantial percentage of its market share, which did decrease from 94% in 1979 to 68% in 1987. During that time, Sprint, MCI, and other smaller long-distance providers concentrated on building a largely fiber-optic platform that took advantage of the regulatory bottleneck that had been put in place.

Nevertheless, the long-distance market suffers from a great deal of over-capacity, and AT&T has been able to halt the erosion of its share in the long-distance market. The viability of further competition in long distance seems limited. In addition, for most of the large companies, competition in long-distance is a bit of a distraction: Success in long-distance does little to directly connect the company to the customer through higher quality or lower cost for local, direct access to telecommunications services.

Companies involved in the long-distance market seem well aware that there is little farther to go in squeezing margins out of services. For that reason, the development of the long-distance platform has largely focused on the addition of strategic assets that provide direct links to the customer and subscriber base. AT&T has made a huge investment in direct customer contact through its purchase of McCaw Cellular, a development that is a focal point for the discussion of wireless access' development in the United States. In addition, AT&T's offering of Internet access in early 1996 to all long-distance customers is, effectively, another entry point into the local loop, even though the service will initially push up usage of RBOC local lines as well.

MCI, the nation's second biggest long-distance carrier, has made investments in the content portion of the platform by putting millions into Rupert Murdock's News Corporation, and, in 1996, launched its bundled service offering, called "MCI One." In 1997, the company turned around and agreed to be purchased by BT—thereby offering an even greater source of cash investment for the company's investment in local and enhanced services.

This need to build a platform that directly accesses local customers is evident in the investment by a number of large telecommunications companies in Competitive Access Providers (CAPs). CAPs basically bypass networks and focus on intensive users of telecommunications services, and they have grown significantly over the past few years, in great part because of the direct investments of various larger telecommunications companies. Because such networks would give local access to interexchange carriers, the CAPs have been positioned as critical parts of the broader service platform of the companies playing a role in their construction. In the case of the long-distance companies, this means more local access and an important extension of their service platform.

There has also been a great deal of investment made in international linkages, which is also a natural outgrowth of the network base for long-distance providers. AT&T has established itself through its Worldsource service, MCI has entered into a strategic alliance with BT, and Sprint linked its assets with France Telecom and Deutches Bundespost Telekom on the creation of yet another global strategic alliance.

From the starting point of a long-distance provider, then, establishing a platform for the future generally means moving back into the business of providing local telecommunications access. Where possible, it also means establishing a global presence which, in turn, increases the value of the platform to a future subscriber base.

Local Telephone Providers (Local Exchange Carriers)

Looking back over the past decade of telecommunications development in the U.S., it is clear that George Zielinski's words from his article in *Public Utilities Fortnightly* in March, 1994 still hold true after the passage of the 1996 Telecommunications Act. "The terms of the divestiture have not prevented the RHC

[RBOCs] from competing, or from positioning themselves to compete further, against each other in the local telecommunications distribution market" (Zielinski, 1994). In fact, even before the passage of the Act, the RBOCs took significant steps in constructing a service platform that would allow them to compete in almost every aspect of the telecommunications and information industry.

The first priority for the RBOCs has been to invest in the platform they already have: the local telephone network. The goal, in advance of competition, is to make that local access network as valuable to customers as possible, with intelligent network features like call waiting, call answering, and the like making up a great deal of the investment in the local networks. When competitors begin to compete with the local platform that the RBOCs have at their disposal, the companies expect to put up strong resistance in their local markets as competition grows. In New York, for instance, NYNEX faced inter-LATA presubscription in 1996, and local callers in New York State were permitted to choose their local phone company. NYNEX followed competitors on a block-by-block basis, countering the marketing efforts of the competitors by touting their capabilities as a complete provider of telecommunication services.

Over the last decade, the RBOCs have become decidedly less local and markedly more "regional" and "national" in their focus. The strategy of "vertical integration" became the pattern for many of the RBOCs as they sought to address the problems of technological change and increasing competition. Each major Baby Bell has been taking stakes in a variety of other telecommunications companies with complementary assets. The strategy was the logical response to the perceived competitive threats of the interexchange carriers and the opportunities that the companies have been granted in cellular and wireless access. As the hype about the information superhighway grew in 1993, the New York Times business section described the Baby Bells as "turning into hunters and prey," fighting to retain their own share of the residential market while exploring a variety of strategic alliances and mergers with various other telecommunications companies.

In the mean time, the RBOCs have applied for and received permission to offer "video dial tone," which would allow them to compete head to head in the delivery of broadcasting and information programming. Their investment in technology and trials, though, have not led any to announce a broad competitive attack on the cable markets—in fact, investments in wireless cable have turned sour for companies like Bell Atlantic and BellSouth which had hoped for a quick entry into that market after the passage of the 1996 Telecommunications Act.

The RBOCs also have attempted to shed their highly fragmented operating structure as well, with each Baby Bell doing away with the nomenclature of the old Bell system's 22 operating companies and presenting one name for the entire company. Although it has mostly been for marketing and public relations purposes, the move to single names and unified corporate cultures took place as part of the downsizing of these companies during the late 1980s and early 1990s. Literally tens of thousands of workers lost their jobs as companies attempted to increase efficiency and improve profitability.

So the RBOCs are building the core of their existing platforms, the local tele-phone networks. They are increasing efficiency where possible, employing new technologies and experimenting with new services, and reorganizing their operat-ing structures. But even with all of these moves and posturing for future competi-tion, the local telephone companies are still hamstrung by the same regulatory principles that have defined their development since divestiture. That is, simply put, because the local telephone network that they are investing in is still regulated under the assumption that the purpose of the RBOC is to provide universal, or near universal service. Price caps are still the issue of the day at the FCC, determining the rates that the local telephone companies charge for access to the local tele-phone network by the interexchange carriers, such as AT&T, MCI and the like. Those charges still account for a very high percentage of overall RBOC revenues, as Table 6.1 indicates. The figures are based on articles from a May 1995 edition of the *Wall Street Journal*.

Table 4.1:
Access fees paid by RBOCs (1994 to 1995)

Company	Access Fee Revenue	Percentage of Total Revenue
BellSouth	$3.9 billion	25%
NYNEX	$3.4 billion	25%
Bell Atlantic	$3.1 billion	24%
US West	$2.7 billion	28%
Ameritech	$2.7 billion	23%
SBC Communications	$2.7 billion	25%
Pacific Telesis	$2.3 billion	25%
BellSouth	$3.9 billion	25%

Note: Figures are from public filings and The Wall Street Journal, May, 1995.

In short, the local platform is valuable, but a great deal of its value is dependent on the interexchange carriers paying access fees. The RBOCs are very aware of this constraint, and have done a great deal to expand their service platform in other significant ways. First, the RBOCs have had much to say about their desire to en-ter the market for long-distance services, but, needless to say, the long-distance companies are not particularly happy about the possibility. In addition, RBOCs have begun to make investments in international transport facilities, such as undersea cables, and local services within a variety of countries.

But, as is the case with the long-distance companies, one of the most significant efforts to increase the reach of the RBOCs predominantly local platform has been through wireless access technologies. The RBOCs have invested heavily in cellular and PCS, opening the door for further growth in a variety of other local markets. In that regard, the RBOCs are likely on a collision course as some of them square off for control over local subscribers.

Given those constraints, then, there are two potential strategies for RBOCs in the wake of the recent legislation. The first is building the platform by scale, and the primary example of this is the recent merger announcement of PacBell and SBC Corporation, formerly Southwestern Bell. In order to combine the scale of local, national, and international operations, while gaining efficiencies through job reductions and forced redundancies, the companies will be combining operations during the course of 1996 and into 1997. Bell Atlantic and NYNEX look to chart the same course as they consider a holding company structure to manage their assets and capital during the period of transition.

On the other hand, RBOCs may choose to build the platform through widening the scope of services, and that appears to be U.S. West's strategy. By choosing to take over Continental Cablevision in early 1996, the RBOC has decided to move away from its rocky alliance with Time-Warner and develop its own assets in the cable markets throughout the nation. The company is now able to offer both cable and telephony, thereby protecting itself in certain markets and giving it a broad range of potential service options as it considers other markets in which it wishes to compete.

Both kinds of strategies are attempts to move around the regulatory and economic constraints that still linger. By building platforms that are less easily undermined by penetration of the local loop by long-distance providers, the RBOCs hope to develop a fortress of strength in the markets they already serve. The initially defensive strategy, appropriate for companies with so much formerly protected revenue to lose, will offer the foundation for a potentially more offensive posture in the years to come.

Cable Television Providers

The cable television industry in the United States finds its foundation in aiding the transmission of broadcast television. As a monopoly service with local franchises under its control, the present platform of the cable companies has largely been configured to support one-way transmission of broadcast information, not the interactive conversations and exchange of information of the local telephone network.

The regulatory push and pull on cable television providers has been highly localized and nationalized at the same time. The issues of zoning and franchise rights have become meshed into the fiber of local politics throughout the nation. National regulation of cable rates, and FCC tracking and oversight of consumer complaints, has put a second form of pressure on the companies. At the same time, cable faces regulation as a "common carrier" of programming, and many cable systems have faced scrutiny for what some allege (and others have attempted to document) as a system for providing unfair advantage to certain programming channels. Yet there is supposed to be "competition" between different broadcast, information, and entertainment options. Like the RBOCs and the long-distance carriers, the platform from which the cable companies begin has been cobbled together in a patchwork of philosophies and regulatory fiefdoms.

But from that starting point, there is much the cable industry plans to do to increase its capability to serve the information and telecommunications needs of customers. Unlike the RBOCs, cable companies have not focused on dramatically upgrading their existing local network, in great part because of cash flow difficulties and a lack of regulatory incentives. At the moment, cable companies are not permitted to provide telephone services in their service area, and the existing plant of coaxial cables used to deliver television programming is of a higher quality than the copper wires of the telephone network.

Therefore, many cable companies instead have concentrated on another portion of their platform as information and telecommunications service providers: the content of programming. The two biggest cable companies, Time-Warner and TCI, have focused on creating cable channels and purchasing the rights to movies and movie production. Another big cable company, Viacom, leveraged itself mightily to purchase Paramount movie studio, after already having developed a host of its own cable channels, such as MTV. With the merger of Capital Cities/ABC with Disney, there has been a great deal of activity in the integration of the content portion of the telecommunications service platform, and cable companies are well aware that their past, as well as their future, is tied to their ability to produce and distribute information and entertainment.

Cable companies do seem to have an interest in moving into the delivery of local telephony, in great part taking the experience that they have had in the U.K. combining cable television with telephony. The mathematics make the move obvious the local phone business is a $90 billion a year market while cable industry's combined annual revenues are about $25 billion. Ironically, though, now that the regulatory constraints have been lifted, companies like TCI have chosen to alter their focus and concentrate on the "core elements of their business"—namely, access to and the development of entertainment services.

One area that cable companies have moved aggressively into is wireless access. As has been the case for the RBOCs and for the local operating companies, wireless investments have not been restricted and as such, form a critical potential extension of these companies' service platform. Through wireless cable and wireless telephony, cable companies are looking to expand both the geographical reach of their services and the nature of their services.

Managing the Development of the Platform:
The Nature of the Challenge in the U.S.

The corporate platform strategies are constrained by tensions between universal service and competition and between local and national autonomy. To gain access to customers, and to begin differentiating kinds of customers according to various market segments, it will be critical to gain direct access to customers and offer seamless communications solutions. This must transcend the local/national divisions and overcome the posed contradiction between universal service and competition.

The regulatory regime is constrained by the same two tensions. On one hand, it would seem a natural companion to the prevalent Washington program emanating from the Republican Congress that local control is better than national control, but the need and drive for economies of scope, if not scale, must be reflected in the economic and political space opened up for activity in the telecommunications and information sector. Additionally, the regulatory function continues to struggle with the perceived political exigency of establishing universal service, even when an opportunity to revolutionize the definition of universal service is evident in the form of wireless access.

Each of these wireline players developed platform strategies in an age where wireless assets were defined and seen as an ancillary service to wireline. The cellular model did not allow for a broader conception of how this access technology could develop. Now the players in each of these category are searching for wireless assets to integrate into their operations, adding to and expanding their service platform so that they will be able to compete effectively for local subscribers. It is significant that wireless access is really the common denominator for all of them, insofar as it is the only access technology that all have identified as a potential mass service for attracting local subscribers. In that regard, the mantra of owning your own platform will be defined, in a significant way, by how each company chooses to operate and manage its wireless network.

How can wireless help to resolve these tensions so that the country can move forward constructively in the development of a national infrastructure that is sustainable and provides the broad range of social and economic benefits that it can make available? Competition in wireless services can be used to cut through the layers of tension and establish a solution for the American telecommunications sector. Before addressing some of the details and the appropriate character of regulation and corporate strategy for wireless access as strategic liberalization, it is important to describe the institutional legacy and framework for the present provision of services in the U.S.

WIRELESS ACCESS IN THE U.S. CONTEXT

Wireless has already clearly been established in the United States as a critical component of the overall growth of the telecommunications sector. As FCC

Chairman Reed Hundt commented in a speech posted on the FCC's website after the initial narrowband auction of PCS licenses in the U.S.: "We expect more than $20 billion to be invested in the immediate pursuit of returns on the auctioned licenses. The result will be more than 300,000 new jobs in the mobile communications business, and another 700,000 new jobs stimulated over the next 5 to 8 years as an indirect effect of this job creation."

To date, wireless access has been able to play a variety of roles in spurring the growth of the country's telecommunications and information sector. From the birth of radio and television broadcasting, wireless communications have been at the center of American life. But, as has been the case around the globe, wireless access as a means for two way communications has generally been relegated to an ancillary role, significantly less important than the copper wires of the public switched telephone network.

As discussed previously, the first major commercial application of wireless communications for two-way telecommunications services involved long-distance and satellite transmission. The growth of competition in the long-distance market is directly tied to the development of point-to-point microwave services, put in place by Microwave Communications International (MCI) and other companies that wished to compete directly with the telephone giant, AT&T. Satellite communications had been developed to support other long-distance and television broadcast companies.

But the use of point-to-point microwave was not to become a mass service, simply because it was not constructed to become so. Wireless local access was limited to geographically isolated regions. Even though cellular systems had been in development for decades, and even though AT&T, the world's largest telephone company, had been one of the pioneers, wireless access in the United States was late in appearing.

It is apparent that the development of cellular services in the United States and the regulatory jostling behind the rollout of new kinds of wireless access services reflects the evolution of the rest of the telecommunications sector. At the same time, wireless is a different kind of creature than the species of wireline access discussed so far in this chapter. Wireless access has been bred in an environment that is disconnected somewhat from the two tensions of national versus local authority and universal service versus competition policy. For that reason, and because of the nature of wireless access as a communications technology, there is an opportunity for the United States to cultivate growth in wireless access to achieve the critical public policy goal of telecommunications development: increased penetration of access to services.

This section begins with the history of cellular access, and turns to examine some of the developments in emerging wireless technologies, such as PCS. The discussion includes a specific case study of how the future of wireless access is determining corporate and regulatory policy today by briefly reviewing the 1994 merger of McCaw Cellular Communications with AT&T, before returning to the themes of sustainable development through a policy of strategic liberalization.

Cellular and the Birth of the Duopoly

The FCC spent more than a decade considering the issue of licensing wireless systems for cellular access. The initial proposal in 1974 was to allocate 40 MHz of spectrum, with one cellular system per market. In great part, the discussion was dominated by the company that would have been permitted to take and operate the licenses under this proposal, namely AT&T. AT&T, through its subsidiary Western Electric and its research arm, Bell Laboratories, had already developed the framework for the provision of wireless access services in the 400 MHz and 800 MHz ranges.

During the 6 years that followed this initial proposal, all sorts of suggestions were offered, including one proposal to share a single 40 MHz license among several licensed users. In the late 1970s, experimental wireless system licenses were granted to Illinois Bell, one of AT&T's operating companies, for the Chicago area, and to American Radio Telephone Service (ARTS) for the Washington and Baltimore metropolitan areas.

After increasing pressure from a variety of industry participants, the FCC finally outlined a framework for the licensing of cellular operators in 1981. The decision was to offer two licenses for each service area, within the service areas defined by the geographical distinctions of Metropolitan Service Areas (MSAs) and Rural Service Areas (RSAs) as defined by the U.S. Department of Commerce.

In articulating the license structure, the potential for anticompetitive behavior was a critical concern of the FCC as it considered the options for determining who could or should not be permitted to participate in the development of cellular services.

> Our primary reason for questioning wireline operation of cellular systems at this late date was our concern that cellular technology might have developed the potential to be competitive with local exchange service, thereby creating a disincentive on the part of wireline carriers to fully develop cellular service in the areas where they also offer local exchange service. From our review of the record, however, there appears to be a consensus that our concern was unfounded. Most commenters believe that cellular systems will initially only be competitive in the traditional 2-way mobile market. Our own evaluation is in agreement with this position. The key to local exchange substitutability in any practical sense is in the availability of an inexpensive handheld portable unit that is light in weight. Until such an inexpensive unit is available, cellular service cannot realistically serve as a meaningful replacement for local wireline exchange service.

> Furthermore, the size of the spectrum allocation will limit the number of users of a cellular system, while a landline system can expand indefinitely.

> Therefore, we conclude that there is no reason to rule wireline carriers ineligible out of concern that they will have a disincentive to advance the development of their cellular system because of its short term potential

replacement of their local landline service. (*An Inquiry Into the Use of Bands 825-845 and 870-890 MHz for Cellular Communications*, CC Docket No. 79-318, April 9, 1981, Federal Communications Commission, p. 16)

The FCC had already discounted the possibility that wireless access could compete with wireline capabilities and, for that reason, felt comfortable with allowing a wireline carrier (namely AT&T) to be permitted to provide wireless access services. This is somewhat perverse logic, in so far as the logic of giving a license to a wireline carrier would seem to indicate the FCC did not wish to sow the seeds for future competition. If such a potential was apparent, why would not the FCC set a structure in place to promote the possibility for direct competition between wireless and wireline service providers? Certainly, if given the opportunity to use a certain model for wireless access as a shield for the massive investment in plant already made by AT&T, the company would promote such a model. To be fair, the FCC did ensure that cross-subsidization between wireline services and cellular services could not occur, through appropriate regulation and the division of operating responsibilities within AT&T. Perhaps the FCC did not think competition in the telecommunications sector as a long-term goal was important, but the prior passage shows that it was not because of ignorance about that possibility.

Another kind of competition was more on the minds of regulators and industry participants, because a few months after the decision was handed down by the FCC, the Modified Final Judgment was announced. The cellular licenses were not at the forefront of people's minds. At the initial press conference explaining the break-up of the company, AT&T representatives did not know whether the new cellular licenses would be kept by the company or by the newly created Regional Bell Operating Companies.

Eventually, that issue was cleared up and the RBOCs were granted the right to operate the wireline cellular licenses within their own service territories. The second licenses were to be granted to "nonwireline" companies, which basically meant any other person or company who wanted to create a company to build and operate such a network.

The FCC decided to hand out the nonwireline licenses through a lottery system, with technical and engineering prerequirements set by the FCC to raise the bar for potential applicants. Literally, hundreds of people who had little or no experience in the telecommunications industry eventually wound up holding licenses throughout the country, forcing companies such as McCaw Cellular Communications to collect licenses and expand their network by purchasing them or buying existing arrangements. The FCC restricted the sale of licenses until the franchises reached the construction-permit stage and actual investments in network capability were being made. But that did not prevent a wild West-type atmosphere among early cellular service providers, with deals being cut and fortunes made and lost on the license speculation.

The first licenses granted were the Metropolitan Service Area licenses, simply because those were the territories where the most evident demand for such services existed. The Rural Service Area licenses were not distributed until 1988, when even more strict guidelines to prevent speculative applications. To this day, though, some RSA licenses have yet to be handed to companies committed to constructing a network to provide cellular services.

As far as technologies go, AMPS became the transmission standard, which had been based on the improved mobile telephone services (IMTS), a trunked radio system. As digital technologies were developed for cellular systems, the debate between the various players has revolved around the transmission standards discussed in the earlier chapters. The major cellular players have diverged on their choice for digital cellular access—companies like AT&T and SBC Corporation have chosen to invest primarily in TDMA systems, while Bell Atlantic/NYNEX and others have opted for CDMA. GSM has even made inroads in the U.S., and companies like Pacific Telesis and BellSouth have opted to integrate GSM into their network upgrade and new construction programs.

The growth numbers confirm that cellular services have been very successful in the American context, with the industry effectively managing compounded growth rates of 20% or more each year since the inception of the service. Cellular, to date, has predominated as the main story of wireless access in the United States, but another story appears to be just beginning: the story of Personal Communications Services (PCS).

The Birth of PCS

The hearings on Personal Communications Services (PCS) opened in December 1991, and was designated as docket 90-314. The docket remained open for the next 5 years as the FCC built to a program for licensing PCS services. During those years, the perceptions of what PCS was, and could be, moved from theoretical speculation to the cold, practical reality of regulatory decision.

The original vision echoed much of the language which appeared in the U.K. as part of the initial "Phones on the Move" report on Personal Communications Networks. Telocator stepped forward with an attempt at a formal definition, which differentiated PCS from cellular services through "personal numbering" and "call management," which were functions not built into existing cellular systems (even though they would soon be, as cellular providers made the transition to digital transmission and switching standards).

The catchall phrase eventually presented by the Commission defined PCS as "radio communications that encompass mobile and ancillary fixed communication services that provide services to individuals and businesses and can be integrated with a variety of competing networks" (FCC, October 22, 1993). PCS eventually was defined as either a private or commercial mobile radio service (FCC, October 8, 1993).

The initial question about the purpose and target audience for PCS services revealed a fundamental division. Some argued that PCN would "evolve from the 40-million-unit analog cordless phone base, for which new digital handsets will be sold that can be used in the streets," (Donaldson, Lufkin, & Jenrette, 1993) very similar to the Telepoint/CT-2 concept that we examine in the British case study. Existing industry participants quite often equated PCS with cellular services. That argument was reflected in the equipment being advocated as the best for the development of PCS networks, which ranged from the CT-2 equipment to the TDMA and CDMA transmission standards under development.

What became clear in the years of debate that followed was that the cellular model was much more powerful than the Telepoint model, perhaps in great part because the failure of Telepoint became more and more evident during the first few years of this decade. By the time the service definitions and licensing structure were set in 1993 and 1994, it was the cellular model that predominated.

Geography was a difficult question. Instead of choosing to follow the MSA/RSA territorial distinction of cellular, the FCC did a surprising about face. The Commission chose to license providers according to the Major Trading Area (MTA) and Basic Trading Area (BTA) distinction provided by the Rand McNally Commercial Atlas. Other than dramatically increasing Rand McNally's sales of this rather expensive atlas, some claimed that there were few benefits of this arrangement, but it held through the proceedings and the 51 MTA areas became the basis of the 30 MHz licenses that were to be granted. License blocks were defined for different parts of the spectrum stretching from 1880 to 2200 MHz, designated A through F. The A and B blocks represented 30 MHz licenses, to be auctioned off in each of the MTA areas. The C through F blocks were smaller and would be auctioned for each of the more than 400 BTA areas.

The choice not to grant national licenses also came under fire, but the decision reflects the tensions between local and national autonomy, as well as the tension between competition and universal service that marks the development of the American telecommunications sector. By fragmenting the market, the FCC attempted to make it difficult to assemble a high concentration of licensees, which would, in theory, meet the needs of further competition.

A further attempt to diversify the marketplace is reflected in the Commission's treatment of existing cellular licensees. The final equation was that cellular licensees were permitted to participate in PCS auctions outside of their service area or in any area where the cellular licensee serves less than 10% of the population of the PCS service area. Cellular licensees were defined as entities which have an ownership interest of 20% or more in a cellular system. Finally, the commission decided to set aside one of the license blocks for small businesses, women- and minority-owned enterprises, and rural telephone companies. As one commentator put it in November, 1994:

> Roughly a third of the bands put on the table have been set aside as "entrepreneurial blocks," theoretically off-limits to fat-cat bidders. In a game in which small is big, the FCC mandated that company bidding on a license

within those blocks must have gross revenues of no more than $125 million a year. In addition, certain bidders huddled beneath the entrepreneurial blocks' umbrella will have the extra edge of being declared "designated entities," roughly the FCC's equivalent of weak golfers who deserve a multistroke handicap. Designated entities are small businesses whose annual revenue is $40 million or less (good for a 10% knockdown on any winning bid) and businesses owned by minorities and/or women (good for a 15% to 25% price cut, depending on annual revenue.)

Sounds complicated? It is. And rest assured that there are fine print rules above and beyond the basic stipulations. Designated entities are not guaranteed winning bids—just a leg up in auction battles with the established cellular giants and Baby Bells of the world. (Dunkel, 1994, p. 37)

Against this continuing backdrop of debate, Congress passed the 1993 Omnibus Budget Reconciliation Act, which called on the FCC to begin auction proceedings by the end of 1994 to ensure that all revenue taken in from the proceedings could be used to reduce the federal budget during that fiscal year. The FCC was therefore compelled to push forward where it could and resolve as many issues as possible to get the bidding process going. By the end of 1994, the auction had begun.

The auction represented a significant innovation in the development of telecommunications regulation and spectrum management, so much so that it deserves some further discussion before we present a brief analysis of the results. The auction might not have been successful in producing the kind of marketplace diversity that some hoped for, but it did set a valuation structure in place for spectrum. This, in turn, opens other opportunities for a program of strategic liberalization to overcome the local/national and competition/universal service tensions that hold back American telecommunications development.

The PCS auction was the first practical implementation of the competitive licensing idea, but the framework and support for such a process had been set through years of discussion and consideration. In 1991, the National Telecommunications and Information Association (NTIA) of the Department of Commerce wrote in its report on spectrum management:

Although changes in regulatory procedures and the block allocation system can improve spectrum management incrementally, the report concludes that greater reliance on market principles in distributing spectrum, particularly in the assignment process, would be a superior way to apportion this scarce resource among competing and often incompatible users. (U.S. Department of Commerce, 1991)

Henry Geller, Assistant Commerce Secretary during the Carter administration, once called the lottery procedure "a national disgrace" in testimony before Congress. The FCC had let more than $84 billion in revenue slip through its fingers.

In short, the lottery did not work for cellular, and something different had to be put in place for any other licensing that was to be done.

An FCC Docket opened to address this possibility of a competitive auction for licenses, and came to the quick conclusion that competitive bidding should begin immediately for Personal Communications Services (PCS), some services regulated by the Private Radio and Common Carrier Bureaus such as the Specialized Mobile Radio, Interactive Video Data Service, and certain cellular radio service applications. The official comments of the FCC indicated that the emphasis in these proceedings was to get smaller, private companies to become part of the auction process, especially companies owned by minority business people and women.

But that is not what happened. Companies began to announce large-scale bidding consortia and mergers of their wireless operations. Sprint came together with three of the country's largest cable operators, including Tele-Communications Inc., Comcast, and Cox Enterprises to form Wireless Co. Bell Atlantic and NYNEX joined forces and merged their cellular assets, only to turn around and come together with AirTouch Communications and U.S. West to establish a four-company partnership called PCS Primeco to bid on licenses. For a while, it appeared that even further consortia would develop, but discussions between MCI and a variety of other large telecommunications concerns did not lead to another alliance. Those alliances, however, were big enough to scare many who were considering their participation in the auction.

The first license auction actually occurred in 1994, for nationwide licenses of narrowband PCS. Almost 1400 licenses were auctioned off, some with 50 KHz and others with only 12.5 KHz. Most of the companies bidding on these licenses were looking to expand existing paging and ESMR operations, rather than attempting to build full-fledged PCS networks.

The broadband PCS auction, representing all of the A and B block licenses in each of the 51 MTA areas, took almost 5 full months to complete, and the stakes were much higher and the players much larger. If the FCC's goal was to bring in smaller players and nontraditional bidders into the fray, the strategy did not succeed, at least for this round of the PCS auctions. Of the 99 licenses sold, 70 went to Bell Companies or Bell Companies in consortia with other auction participants. Wireless Co., the Sprint-led consortium, was the biggest winner, with 29 licenses totalling more than $2.1 billion. AT&T purchased 21 licenses for $1.6 billion, and PCS Primeco, the partnership of Baby Bells and AirTouch took 11 licenses for $1.1 billion.

From the perspective of income for the treasury and debt reduction, the auction has turned out to be reasonably successful, with a total of more than $7 billion coming into the U.S. Treasury. But, on the surface, the auctions did not diversify the market for wireless access, insofar as the players in the auction are all major participants in the cellular access market and have construed their services largely along the lines of cellular access. For example, AT&T expects to have its PCS service running by the first half of 1997, and will integrate the new PCS services with existing cellular services to provide seamless wireless communications capabilities, utilizing dual-band, dual-mode wireless phones.

Since the completion of the broadband PCS block A and B auctions in 1995, the FCC has launched a variety of auctions for a wide range of wireless access services—many of them potential competitors to the new PCS licensees. In January 1996, the FCC auctioned two permits for Direct Broadcast Satellite (DBS) services. One of the permits offered full nationwide coverage, while the second covered most of the U.S. but not the eastern coast. The auction prices reflected the higher quality of the first license, which went for more than $680 million, as compared to the second, which sold for a little more than $50 million.

The PCS C-Block auction took place in early 1996, with dozens of bidders competing for BTA apportioned licenses throughout the U.S. NextWave Personal Communications, Inc. was the most aggressive bidder, spending more than $4 billion on licenses in New York, Los Angeles, Washington, Boston, and a variety of other cities throughout the U.S.. Questions have been raised, though, about the viability of these newest licensees—in great part because a huge amount of capital has just been expended for purchasing licenses which cannot be used for infrastructure and service provision.

Following quickly after the C-Block auction, the FCC launched a competitive bidding process for Multipoint Distribution System (MDS) licenses, also on an BTA basis. MDS, sometimes referred to as wireless cable, has been a item of great interest to both existing cable providers and incumbent telcos like Pacific Telesis and Bell Atlantic. The interest of large and small players is represented in the final results. Pacific Telesis was the second largest bidder, behind CAI Wireless Systems. The total auction proceeds were just over $200 million—less than the PCS auctions, but still a significant contribution to the national coffers.

Looking to the future, the FCC recently adopted rules for a new category of wireless services, called "General Wireless Communications Service" (GWCS). Twenty-five MHz of spectrum in the 4660 to 4685 band will be transferred from the federal government to the private sector, and the FCC specifically adopted competitive bidding rules for awarding mutually exclusive licenses. So it looks likely that the auctions will continue for the foreseeable future, at least until the FCC's auction authority sunsets in near the end of the century.

Pioneer's Preference

Another critical innovation developed as part of the proceedings on PCS licensing was the establishment of a "pioneer's preference" program. The pioneer's preference was established by the FCC to reward companies for innovations in wireless access technology or service development. Conceptually, the idea makes much sense and has a great deal of merit: Those who work to improve the opportunities to provide telecommunications services should be rewarded, when possible, for their efforts and investment. In that way, the pioneer's preference forms another potential economic incentive that promotes innovation in the development of the telecommunications sector. But the theory was much cleaner than the practice. The whole process for awarding pioneer's preferences, and the establishment of rewards, fell into the morass of telecommunications politics in Washington.

The first official statement about such a proposal appeared in 1990, when the FCC linked technological innovation to requests for the spectrum required to provide a new service. The FCC's General Docket 90-217 Pioneer's Preference Rules allowed preferential treatment in the licensing process to parties that "develop significant new communications services or technologies." In this regard, the key words are "new," "innovative," and "original." Chairman Alfred Sikes later explained that there must be a "steep winnowing process when the principal criterion is originality" (Telecommunications Reports, 1992).

The incentive for being "original" formed the basis of what was to become an extended political and legal battle. Not surprisingly, a number of the nation's telecommunications companies lined up to be considered for the pioneer's preference, claiming "originality" in their ability to provide wireless services. Among those was AT&T, which requested an exclusive pioneer's preference. Considering that the goal of the program was, in part, to reward innovation by companies trying to break into the market, the idea of awarding such a license to AT&T, or to many of the other established players who filed for consideration, would certainly have defeated the purpose.

In October 1992, the FCC awarded three pioneer's preferences to American Personal Communications (APC), Cox Enterprises, and Omnipoint Communications, and denied 53 other requests for similar treatment. APC's innovation was the company's Frequency Agile Sharing Technique (FAST), which optimized PCS channels, independent of both modulation and access techniques. Cox was granted a preference because of its development of technology to use cable television plant for connecting PCS microcells. Omnipoint had completed innovative work on spread-spectrum techniques for transmission of information.

Having discriminated between winners and losers, the FCC now had to determine exactly what the winners had won. In the initial formulation, the pioneer would immediately receive a license to provide the new service, with no competing licenses to be awarded for six months (Kagan, 1994). But that idea was derailed with the rise of the Congressional debate on the passage of the General Agreement on Trade and Tariffs (GATT) treaty in late 1993.

Legislation passed to implement the GATT used the spectrum valuation of the PCS auction to determine the exact nature of the reward. Pioneer's preference winners must pay 85% of the average price paid for comparable licenses sold at auction. The amount can be paid in a lump sum or in installments over a 5-year period. The FCC was granted the right to identify comparable licenses and apply the payment formula. On completion of the auction, APC chose the license for the Washington and Baltimore market; Cox Enterprises won a license for Los Angeles and San Diego; and Omnipoint received a New York license. The three had to pay a total combined price of $700 million for their licenses.

The program's effectiveness has been questioned, even by those who have been granted the licenses. The politics behind the identification of pioneers and the relative merit of new technologies is difficult to assess, even in the least confrontational and competitively charged environments. Legal challenges continue in the courts and likely have killed any possible thought of resurrecting the program for use in future auctions. In addition, the FCC's right to offer pioneer's preferences ends on September 30, 1998.

Even so, the program represents an important first step for a regulatory authority in the telecommunications industry—the pioneer's preference program represents perhaps the first major attempt to play a proactive and facilitating role in the development of the telecommunications marketplace. In addition, three companies that might not have been able to play a role in the development of Personal Communications Systems found a place at the table and could participate in the auction process. Looking at the program more broadly, one could even claim that the combination of auctions and preference programs offers a clear economic incentive for research and development of new technologies; with a 15% discount offered on licenses after auctions are completed, there is a definite incentive to win.

Wireless Access and Convergence in the U.S. Telecommunications Industry

The prime case study of the potential for the convergence of wireline and wireless access comes from the giant of U.S. telecommunications—AT&T. In reviewing the purpose and execution of AT&T's purchase of McCaw Cellular Communications in 1994, we begin not only to understand how the present of wireless access has been defined by the past, but also point to the best future for the development of this kind of corporate institution as it relates to our overarching public policy goal of increased and enhanced service penetration.

AT&T's relationship with McCaw Cellular Communications began in 1992 with a purchase of a 19% ownership stake in the company. Early on, the move to take a stake in the company was seen as an opportunity to use wireless to break into the local loop:

> AT&T's purchase of a stake in McCaw was not a total surprise. For AT&T the stake in McCaw/LIN has both offensive and defensive attributes. First, it enables AT&T to influence—and perhaps someday control—the largest cellular operator in the United States. Moreover, most of McCaw/LIN's pops are in major cities (including New York, Los Angeles, Dallas, and Miami), and roughly two-thirds of all domestic long distance calls originate or terminate in the top 20 to 25 U.S. cities. Second, we have long hypothesized that if roughly 20% of the U.S. population is going to use or own a wireless handset early in the next century, heavy communications users (people with a strong need to be continuously in touch) will constitute the vast majority of cellular subscribers. We would further argue that

the heavy cellular user is also likely to be a heavy user of long-distance services. The combination of these two observations means that AT&T had to get in front of that potentially huge long-distance calling pattern and subscriber base of the future, or else risk losing it to a competitor like MCI, Sprint, or some other party acting as a "wholesale" long distance carrier that McCaw could market on its own as "McCaw long distance."

From an offensive perspective, the McCaw deal will enable AT&T to better utilize its brand name and distribution channels; to sell cellular infrastructure equipment to McCaw/LIN; to sell consumer and business wireless hardware; and to tap the potentially huge emerging market for wireless data and messaging. It probably improved AT&T's visibility and status in bidding for the new international wireless licenses. (Donaldson, Lufkin, & Jenrette, 1993, p. 12)

This is certainly a definite vote of confidence from a prestigious investment house, but it also reflects what was to be perceived as a strategic necessity for AT&T: gaining access to local customers by integrating its access offerings. This was a business that AT&T invented and abandoned more than a decade before, and now it was trying to rebuild its position in that business. Eventually, AT&T decided to purchase the whole company, a move which cost $12.6 billion.

Why did Craig McCaw sell out the remainder of the company? As Bob Ratcliffe, McCaw's Vice President for Corporate Communications in 1993, said, "He was looking for a new challenge. And he had taken it about as far as it could go." Of course, the incentive of about $1 billion in AT&T's stock certainly did not hurt matters much, either.

But there was likely a deeper issue. Despite the tremendous growth of cellular, and the projections for the next wave of PCS growth, McCaw Communications never really made money. In 1992 alone, McCaw Cellular lost $364 million, in great part because of the mass of debt that had been built up during the explosion of growth in the 1980s.

AT&T has a great challenge ahead, one that faces each of the wireline service providers with wireless assets: how to merge together the wireline and wireless businesses so that the company can directly deal with the customer. AT&T faces a challenge in bring the two cultures together; AT&T is known as a very hierarchical organization, while McCaw's culture is characterized by words like "chaos" and "frenzy."

There were also regulatory hurdles to overcome. After AT&T purchased McCaw, the company was restricted from using the AT&T name to market wireless services for one full year. As soon as that restriction ended, AT&T unveiled joint wireless and interexchange offerings. The aggressive move by the company came, in great part, because of MCI's purchase of a long-distance cellular reselling company, which brings MCI directly into competition with the AT&T/McCaw service network in providing bundled long-distance discounts to cellular subscribers. Now with AT&T's "third divestiture," the push to integrate the structure and marketing of the company's integrated services will likely receive even more life and energy.

What it came down to was this: Wireless access needed wireline access, and vice versa. The financial burden of building a wireless access network was a critical factor, but it is the economy of providing a single platform to customers that makes wireless access a critical component to any company's integrated platform. By purchasing McCaw, AT&T immediately became the largest wireless service provider in the country—and AT&T expanded its lead in potential customers served through its subsequent aggressive bidding at the PCS auctions. As Table 4.2 indicates, AT&T is potentially the dominant integrated provider of wireless and wireline services in the country.

Considering the efforts of some to spin off wireless franchises and licenses, this claim is certainly not the consensus view in the American industry. But if McCaw Cellular cannot push forward, even when it is at its strongest and the margins are thickest, how can wireless survive without wireline in an environment with razor-thin margins and potentially dozens of competitors? That is the question AT&T has to ask as it looks to move more directly into the local loop and position itself aggressively in the wake of the passage of the 1996 Communications Act.

So, how does this kind of strategy fit into the continuing structural change in the U.S. telecommunications industry, as well as the potential role of wireless access in the evolution of the competitive marketplace? In part, it goes back to the issue of owning a platform and building a competitive presence that is based on not only a set of discrete products and services, but also the technological infrastructure required to support it. AT&T, as is the case with all long distance companies, is betting on the convergence of wireless and wireline systems into the broadly ubiquitous platform that people like Calhoun and others point to as a competitive necessity.

It also ties in with the two primary fissures already identified: national versus local authority and universal service versus competition. Wireless as an access point for the local loop in the U.S. context benefits from the cover that it is given as a service largely regulated at the national level—effectively, by providing local access through wireless services, AT&T could get around the difficult political game within each state jurisdiction. Although AT&T does not have specific universal service obligations, wireless also offers the opportunity to position the company again as a "universal provider of services," if not the only telephone company and the object of regulator expectations in specific markets. So, from the perspective of positioning a corporate strategy in a difficult regulatory environment, wireless is a safe way to get into the local loop without tripping over the stones and pitfalls inherent in the U.S. system of telecommunications provision.

Table 4.2:

Top ten cellular/PCS providers (1995)

Company	Cellular Pops (thousands)	PCS Pops (thousands)	Total (thousands)
AT&T	78,562	111,821	190,383
AirTouch/ US West	63,236	14,403	77,639
Bell Atlantic/ NYNEX	54,310	1,752	56,062
GTE	54,202	9,665	62,867
BellSouth	40,433	12,226	52,659
SBC Corporation	38,729	40,474	79,203
360 Communications	28,759	0	28,759
US Cellular	24,184	0	28,184
Ameritech	22,033	8,195	30,228
ALLTEL	8,261	0	8,261

Note: Public filings by providers and Donald, Lufkin & Jenrette (1996)

And then there are the results of the auction and the pioneer's preference program to consider. Although this is not the place to offer an in-depth examination of the auction results, it is clear that the auctions have provided a great deal of leverage to existing cellular providers. The broadband PCS auction has allowed the major cellular players to protect the investment they have already made in cellular services. If companies do plan to roll out cellular-type services (and provide cellular-type pricing), penetration rates will stall and the move to replace the landline phone through direct competition will also fail. The broadbased competition which wireless could potentially bring would have failed, one could argue, in large part because of the auction and the nature of the participation.

It is these smaller licenses, this swarm of potential confusion, that could be more viable over the long term. As service providers launch offerings, fail, reorganize, merge, and divide, different kinds of services are likely to become viable on a variety of different licenses in the spectrum. Then there will be fascinating opportunities for niche and broad-based competition, perhaps even providing enough support for the kind of "telecommunications biodiversity" which would be appropriate for a country as broad and diverse as the U.S. If the auction has been a frustrating necessity to motivate government to action, then this kind of conclusion may mark the events of the past 2 years as a necessary evil on the path to a healthier telecommunications industry.

Finally, it is important to point out that the auction and any incentive programs connected to it is a step to a final goal, and cannot in and of itself provide the framework for competition. Policymakers need to move the focus of corporate strategy from competition for licenses to competition for customers—and a competition to build broadly competing service platforms that are meant to provide access to all of this country's citizens. This is an appropriate note to sound as the discussion turns from the descriptive to the proscriptive to focus on the various infrastructure development theories and their lessons for corporate strategy and the character and level of regulation in the U.S. telecommunications market.

STRATEGIC LIBERALIZATION AND INFRASTRUCTURE DEVELOPMENT IN THE U.S.

The models for wireless service provision in play at the moment in the U.S. are largely based on cellular—even though PCS auctions have been completed and there are a variety of potential opportunities arising from the completion of the auctions, few would argue that the reality of the market today is far from the vision of a competitive market based on true, facilities-based competition. In that regard, the pattern of infrastructure development in the U.S. is hampered by the constraints imposed on service providers—not only by specific kinds of regulations, but also by the endless tendency of the corporate environment to focus on replicating past successes and well-work business plans.

Infrastructure development programs based on the ideas of cultural and technological protection, or on the theories of market subsidization, would see little problem in this. Cellular services are, essentially, ancillary and aimed at the higher income groups of the developed and developing world. As such, wireless access is something to be criticized, or referred to in footnotes as pulling resources from the main telecommunications networks that serve socially significant purposes. Although that may certainly be a justified position, and one that has been echoed in earlier portions of this discussion, it still does not offer much positive direction in building a different kind of wireless service model appropriate to the specific history of the U.S. context.

On the other side of the political spectrum, the techno-libertarians and strong liberalization advocates have programs that would revolutionize the whole of the telecommunications markets, but that sometimes do not take into account the path-dependent course of telecommunications development. In the U.S. case, it is clear that the divisions between local and national, and between competition and universal service, make many of the suggestions which come from these camps difficult to implement and unpalatable to a variety of groups. Perhaps this is a good thing—especially in a country where one of the founders advocated revolution every now and again as a course of policy.

But there is still the fact that wireless is often grouped as one of many different areas of potential deregulation. The terms of the debate are set so that the potential future value of wireless access is discounted because of its present position as an ancillary, secondary service to wireline. As the cost numbers shift and more and more consumers see wireless as a potential alternative, these kinds of infrastructure development strategies will be challenged to look farther than the broad strokes and provide a more detailed roadmap of new service models for wireless access.

This is also the area in which the regulatory mainstream falls short. The regulatory mainstream tends to focus on the details of one kind of regulation over another and the importance of specific licensing policies, often missing the forest for the trees. Although it is all well and good to point out the structural similarities and differences in the various approaches, at the end of the day, it is necessary to speak to more than just the descriptive and offer some alternative course.

Wireless access, as the basis for a broadly comprehensive infrastructure development program for the U.S., offers specific advantages, given the history and competitive context of the U.S. Only wireless access can be local and national at the same time. The cost of building a ubiquitous wireline network, a truly national network, is prohibitive. It will not be possible for a company to develop a platform that competes nationally without some major investment in wireless access. Even if all of the Baby Bells were to merge together tomorrow, it would only be a temporary solution; eventually, wireless service providers would acquire the advantage of lower variable costs in maintenance and lower fixed costs in infrastructure investment, and undermine the viability of the wireline network.

Only wireless access can be universal and competitive at the same time. The whole idea of universal service is revolutionized through wireless services: It is no longer universal service, but universal access that is important. As soon as everyone can access, for a reasonable price, basic telecommunications and information services, then the true public policy question can be asked: The private sector has offered access; how can we help people pay for access? That question is not for this discussion, but setting the framework so that the question can be asked and answered by government and corporate managers in cooperation certainly is.

Tying wireless access to the dominant character and trends of the U.S. marketplace requires a focus on the character of intervention in the telecommunications market, as well as the substance of government policy and corporate strategy. The remainder of this chapter takes each of these issues in turn and elaborates some of the themes that are the basis for comparison in the remaining case-study chapters, as well as in the conclusion.

The Level of Government Intervention

Identifying the appropriate level of government regulation in the U.S. context can be a difficult task, if only because the various overlapping local and national jurisdictions make it difficult to identify whose regulation applies to which part. The passage of the 1996 Telecommunications Act is likely to make things more complicated before they become simpler; even though there is much talk of the FCC and other federal agencies slowly asserting more and more authority over the state PUCs and PSCs, the "federalization" of telecommunications policy, if it is to happen at all, is a process that will stretch over decades, not just years.

The level of government intervention in wireless services markets, though, is almost entirely a national concern, and it greatly simplifies the task of mapping out an appropriate foundation for an infrastructure development policy when it is based on the capabilities of wireless access. Even so, the tight control over aspects of the wireless service markets in the U.S. represents potential difficulties to the evolution and growth of the industry. Two particular areas of government regulatory prerogatives should be reconsidered in the U.S. as part of a strategic liberalization approach: license definitions and seamless service provision (i.e., both wireline and wireless access).

License Definitions—Stepping Back From Unnecessary
Regulatory Distinctions

For wireless access to become national, local, and universally available, wireline and wireless capabilities will have to be merged. But before wireline and wireless accesses converge, there has to be a regulatory move to promote the convergence of the various kinds of wireless access.

Again, the discussion starts from the well-established point that bandwidth is bandwidth. Any services that are provided on one part of the electromagnetic spectrum can be supported elsewhere, albeit with different technical specifications and equipment modifications. Given that fact, what is to prevent a licensee for wireless cable from providing PCS-type services? From the technical point of view, absolutely nothing. From the regulatory point of view, everything is structured to prevent an operator from crossing over service boundaries, even though changes put in place by the FCC at the beginning of 1997 do take the first tentative steps in the direction of increased flexibility in service and license definitions.

But even these changes only touch on the edges rather than striking at the heart of how the regulatory institutions view wireless licenses. As George Gilder put it in a June 1994 *Forbes ASAP* article: "It is as if the FCC's Reed Hundt is auctioning off beachfront property, with a long list of codicils and regulations and restrictive covenants, while the tide pours in around him and creates new surf everywhere." Gilder's view is that the law of the microcosm will push network intelligence to the edge, replacing the dumb terminals in the traditional cellular paradigm to the end users, and that the present framework for wireless access resembles the mainframe systems which were blown apart by personal computing in the 1980s. So the constraints of licenses will not hold back technology, only impede its progress and limit the development of the telecommunications sector. The question is, therefore, how do we get beyond licenses in the American context?

Licenses are defined according to the kinds of services that can be provided. But let's be blunt: Given the kind of strategies expressed for PCS during the build-up to the auction, is there really any difference between cellular and PCS offerings? No, that's openly admitted by the industry participants. So why is there the illusion that "cellular" is in one band and "PCS" is in another band?

Another example: When the FCC begins to license interactive wireless data licenses, what will be the difference between the services provided there and sending a fax over a wireless cellular network using CDPD? Technically, nothing, it's an image, encoded and sent through the air from one place to another.

Wireless services must be permitted to converge, if not in name, then at least in character. Regulatory policy, and corporate strategy, needs to move toward an understanding of the basic common denominator of wireless access and begin to permit any services to be offered on any bandwidth.

Part of the purpose in creating such divisions, the regulators would say, is to help establish a structure for competition. By creating the distinction between PCS and cellular, for instance, it was possible for the FCC to forbid cellular providers to purchase and stockpile bandwidth during the auction proceedings. But what prevents the FCC from establishing an overall limit, say 50 MHz throughout the whole spectrum within any specific geographical area, to sustain that position? Now that the Baby Bells have begun to purchase wireless cable franchises, that number will probably be too low, and perhaps the FCC might get put in a position where companies have bits of all sorts of spectrum and the Commission will not be able to do anything about it post facto.

In other words, deregulation in the U.S. context implies the removal of license restrictions on the kinds of wireless services that providers can make available. Existing license structures cause many companies to divide their services according to specific parts of the spectrum, even though the same kinds of services can be offered in different spectrum blocks. Broadcasters, who have access to a sufficient amount of spectrum to continue sending television signals and run a cellular company or two within their service area, can not consider the launch of such services. Until these artificial and outdated regulatory divisions are torn up, service providers will not be able to provide the kinds of solutions that many customers want. By pushing wireless convergence, the long term goal of platform convergence will be facilitated, thereby enabling further competition and increased service penetration.

Seamless Service Provision

The issue of cross-subsidization and concentration of commercial power are rightfully at the center of regulator's considerations as they look at wireless access services. In great part, the history of wireless access has been built from the fear that wireline providers will use wireless services to subsidize their wireline offerings to the competitive detriment of other companies. The tight regulation is reflected in the structure of the major providers in the U.S.—every Regional Bell Operating Company has a nonregulated subsidiary responsible for wireless service provision. In some cases, those subsidiaries are directly tied to assets of other companies that may also become future competitors for local wireline services.

Wireline service providers have, because of regulations, developed a "holding company" mentality when it comes to wireless assets; they are investments which should be gauged according to their independent returns, rather than assets that, when directly integrated with the wireline network, provide synergies and add value to overall service provision. That, in turn, perpetuates the impression in the U.S. that cellular is just a subsidiary service that should remain a second class citizen to broadband access technologies.

This is another area where decreased regulation will be required in order to break the cycle that has descended on U.S. service providers. Barriers to cross-subsidization, and restrictions on the nature of company holdings, will have to continually decline as competition increases, and particular attention needs to be paid in order to ensure that the walls between wireline company operations and wireless subsidiaries are taken down as quickly as possible.

The problem, in the U.S. context, is that many of the decisions on cross-subsidization and investment policies for the telecommunications providers take place at the state level. This is an area where preemption will have to become a priority if state governments are unwilling to accept the development of increased competition in the local loop and in the wireless services market.

In sum, regulators should let companies provide seamless wireline and wireless access. In most of the world, regulators have compelled the bifurcation of wireline and wireless services to ensure no cross-subsidization of wireless services by wireline revenues takes place. This regulatory step, although necessary in less competitive times, now threatens to severely hamper the ability of telecommunications providers to offer total solutions to individual customers.

The Substance of Government Policy

Beyond the reduction of the regulatory purview in the U.S. context, there are a variety of substantive policy decisions that have been reviewed in the discussion so far that deserve a second look. The priorities of government policy, when tied to the overall objectives of strategic liberalization set out in the opening chapters, should be oriented to facilitating the evolution and sustainable development of the wireless services market. As such, the discussion in this section is confined to some of the issues that have been identified as most critical to the substance of government policy in the U.S. Two, in particular, are of interest: co-location of wireless facilities and the role for auctions and incentives.

Co-location of Facilities

To run a wireless network, sites for cells and transmitters are required. Once the licensing is done, the most difficult part of constructing such a network is finding the most suitable sites for locating cells. And there are two dimensions to finding a suitable site: The sites have to be technologically suitable (i.e., positioned to provide maximum coverage in appropriate areas), and politically suitable (i.e., acceptable to local authorities with control over zoning issues).

Both issues directly impact the costs of constructing and operating a commercial mobile radio service. For example, it is estimated that costs involved in the acquisition of cell sites can account for about one fifth of the cost associated with making a PCS cell operational. That would include not only the purchase of the actual site, but the research and occasional political arm-twisting required to get things done at the local level.

As is the case with other areas of conflict between local authority and national development, the problem of zoning and location of facilities is complicated by the vagueness of law and regulation:

> The issue of federal pre-emption of local zoning and other regulations represents a battle between two valued, but conflicting, public policy goals. On the one side, federal policy makers, as set forth in the Communications Act of 1934, are trying to bring advanced communications services to the public. On the other side, communities and citizens are trying to preserve local control over their land and affairs—a long-standing tenet of American political culture. In essence, the issues surrounding federal pre-emption of local regulations affecting antenna siting derive from ambiguous language contained in the omnibus Budget Reconciliation Act of 1993—the legislation that established the Commercial Mobile Radio Service (CMRS). In that Act, Congress stated in part [that] "no State or local government shall have any authority to regulate the entry of or the rates charged by any commercial mobile service or any private mobile service, except that this paragraph shall not prohibit a State from regulating the other terms and conditions of commercial mobile services." Each side in the pre-emption debate has interpreted this passage as supporting its position. Without adding information or clarification, congressional intent regarding pre-emption in the case of zoning and antenna siting remains unknown. (Office of Technology Assessment, 1995, p. 202)

Besides providing further support for our ongoing discussion of local versus national authority as a defining fact of U.S. telecommunications development, that quote makes it clear that Federal action to preempt local authority on this issue is likely to be problematic, even in light of the passage of the 1996 Act. But one of the tenets of the strategic liberalization policy could play a significant role in alleviating, at least temporarily, this difficulty. The co-location of wireless facilities has already been shown to enhance the ability of service providers to reduce the costs of building and running a network by lowering the price of site locations. Interestingly enough, co-location might also be the answer to the problem of local versus national authority on cell locations.

By aggressively pursuing the co-location of facilities, either by requiring such agreements to be made or facilitating agreements through appropriate economic incentives, the cell sites presently in use today for cellular services are much more likely to become available for the expansion of newly launched PCS services. Granted, this is only a temporary solution; PCS licenses will require more cell sites in order to ensure maximum coverage, so replicating the site map for cellular would not be enough. But at least this would give PCS providers a critical first step in putting their network together.

Pursuing this policy option has the added advantage of bringing down provider costs as well which, in the long run, is much more likely to increase the penetration of services. Although co-location will not resolve the issues facing regulators and Congress, it provides an opportunity to circumvent those issues and get on with the business at hand: increasing the availability of wireless access services to America's citizens.

A Role for Auctions and Incentives

With the reality of convergence between wireline and wireless access becoming more apparent with each passing day, the process for auctions and pioneer's preferences becomes clearer and more efficient. Auctions are no longer about one particular service, but focus on opening up the market to new participants. After the market cannot bear any further entry and the sufficiency of spectrum is no longer a barrier to entry, the regulators will have achieved their most critical goal: They will have put themselves out of a job.

But there are a number of steps that need to be taken before such a possibility could even be conceived in practice. The FCC still has much to do in ensuring opportunity for new market entrants and motivating technological innovation during this period of intense technological change. That is why the pioneer's preference program, with all of its troubles and evident pitfalls, might have a future role.

What has not generally been discussed in the trade press is the necessary connection between the auctions and the pioneer's preference program, a connection that was not recognized by the FCC until late in the game and was eventually imposed on the Commission by Congressional dictate. The auctions set the ground for incentives to be established by regulators that will aid in achieving the critical public policy goal of increased penetration and enhanced service quality.

Spectrum licenses have been valued through the auction process. If the FCC chooses (and can make the appropriate political arrangements), it can announce that licenses will be available for wireless services in any of the nation's MTA markets, say a 30 MHz license in the space just above 2 GHz. The license would be given to any company willing to apply for the award at 85% of the cost of the license, and the winner of the award could choose the service territory of its choice. Slowly but surely, companies would target innovations and opportunities to fill gaps in their networks, or small companies would continue to focus their work so that they could take advantage of the program. The cost in administering the program would not be too substantial, and the criterion for acceptance of applicants has already been set. The only downside for this program would be that in the areas where awardees choose to provide new services, the level of competition would be increased. But, in a competitive market with no barriers to entry, that is just a fact of life.

This connection between auction processes and regulatory incentives deserves to be explored further, but there is definitely a lesson to be learned at least from the discussion. Strategic investments in wireless services can be harnessed to induce further competition in the market, because such an investment can be tied to reductions in barriers to entry. And, in an industry which has been characterized as a natural monopoly because of the high fixed costs of entry, this is the most critical struggle in establishing a competitive environment.

The Direction of Corporate Strategy

In a highly regulated environment, corporate strategy depends a great deal on the cues sent from the government policy maker to the corporate manager. In a competitive environment, the direction is supposed to be set by the corporate manager with government response when needed for purposes of protecting or promoting the public interest. Corporate strategy in the U.S. telecommunications industry needs to move from the former characterization to the latter; the discussion has made it clear that corporate strategy still remains captive to the interests of local and state regulators, and is likely to remain so even if full, facilities-based competition becomes a reality much sooner than has been projected in this book.

The balancing act for the corporate strategist looking at the U.S. wireless market, therefore, is to balance the direction of government policy with the opportunities opened by the penetration of new technology and the realization that wireless access will be able to compete with fixed services in the foreseeable future. The specific goals of strategy are generally tied to the specific interests and positions of particular companies, but it is possible to generalize and address two areas of concern which any forward looking strategy would have to address: fixed wireless services, along with outsourcing and reselling.

Fixed Wireless Services: An Opportunity Rediscovered?

The predominant model for wireless access service has been cellular access, a model that suffers from the liability of being a premium service in a world requiring cheaper forms of wireless access. Considering the American romance with the automobile, and the fact that most Americans commute to work every day in an automobile, the fact that cellular systems have relied on providing high mobility to customers comes as no surprise. But as companies integrate platforms and begin to see opportunities to reduce costs and gain a greater market share, the possibilities of moving away from a cellular model will become more and more attractive.

The clearest alternative is the fixed wireless local loop architecture, which provides wireless access between the telecommunications network and a fixed point, such as a home or office. Such systems are less expensive than those used for cellular service provision, and the difference in pricing would allow a market entrant to undercut the prices of incumbent carriers if it could make the service attractive to users.

There is some experience in the United States with fixed wireless local loop applications. Perhaps the most famous is the Basic Exchange Telecommunications Radio System (BETRS). The FCC permitted the new class of radio communications to allow companies to use digital radio instead of copper wire in situations where it was more cost effective, which meant mostly rural applications. By the early 1990s, there were more than 50 BETRS systems serving several thousand customers (Calhoun, 1992).

Other experiments with wireless local loop applications in the United States have primarily been based on the same technology as appeared in the U.K. for CT-2. In 1989, Cellular 21 became the first company licensed for CT-2 experiments in the U.S., and they were permitted to use the 866 to 868 MHz band allocated usually to regional public safety communications. As of April 1993, there were more than 200 license applications for PCN/CT-2 technology trials from companies like AT&T, GTE, Ericsson, Motorola, and a variety of other large and small companies. Certainly, there was at least a broad-based consideration of the way wireless local loop technology might apply to the competitive marketplace.

But the cellular model won the day in the run-up to the licensing of PCS. Many of the initial entrants will use PCS as another way to provide cellular services, emphasizing mobility and the functionality of its all-digital network. But as companies face increasing competition and margins are cut thin, there will be compelling reasons to turn to fixed wireless local loop applications as a product differentiation scheme.

For example: AT&T decided in early 1997 that it would use its wireless assets to directly compete in the local service market. AT&T Wireless Services and AT&T Labs jointly developed a proprietary fixed wireless system that will provide two phone lines and capability for Internet access at 128 kbps via a transceiver the size of a pizza box mounted on the side of a house. That transceiver will transmit to a neighborhood antenna capable of serving up to 2,000 homes.

The local, fixed wireless service was built to capitalize on the 21 broadband personal communications service (PCS) licenses acquired in March 1996, in addition to the 222 additional licenses the company won during the D-, E- and F-block PCS licenses. All totaled, these licenses cover 93 percent of the U.S. and give the company access to a sizable customer base. Testing for the service begins in 1998, with rollout scheduled for later in that year.

This kind of market development scheme is the first step in replacing the land-line phone, which is exactly what AT&T will have to do if it wants to establish a universal network and offer services to everyone. Because the future is in competing networks, George Calhoun's vision would be achieved: Companies would be using a variety of wireless access schemes to promote a world where access to the local telecommunications networks is carried through the portals of the airwaves.

Outsourcing and Reselling

The trends of downsizing and "rightsizing" in the U.S. have not encroached on the discussion so far, but the impact is clear in the statistics on employment levels in the telecommunications industry and the stated corporate goals of achieving enhanced shareholder value through cost efficiencies. Two of the elements often discussed in the creation of agile and competitive institutions are outsourcing and reselling. Both are considered to offer competitive advantage when used appropriately. The preliminary discussions of the theory of strategic liberalization indicated that both offered great promise, especially in developing world contexts where resources are always limited and managerial capabilities often scarce.

For the U.S., though, outsourcing and reselling may have less significant roles. The areas where economies of scale may be possible (such as billing, marketing, and the like) are often considered areas of competitive advantage for a corporate strategist. It is doubtful that a number of companies would want to share billing systems, for example, if it were at all possible that billing information would be compromised and shared with other companies. One company outsourcing its billing to a vendor may be appropriate but, from a economic point of view, the cost efficiencies represented are only those of the reduced cost to the client, rather than those achieved through the economies of scope associated with two or three companies joining together.

In the U.S., the Bell Operating Companies have had the pleasure of getting the bills to customers and administrating all of the paperwork required for both local and long distance. Someday, these companies are likely to want to rid themselves of the burden and concentrate on providing clear and regular billing statements that are somewhat different from those of other competing companies, if only for the purpose of differentiating themselves in the marketplace.

For wireless service provision, billing systems are extremely difficult, especially when the problem of security is added into the mix. The network needs to support all of the roaming features, calculate the air time and landline charges, keep track of the service plans—most of this is a matter of software but, certainly, the software becomes the source of the competitive advantage insofar as it embodies the pricing plans and the network's capabilities. If common outsourcing for billing is unlikely for wireline service providers, it is even less likely for wireless providers.

Reselling has been an important part of allocating excess capacity in local and long-distance markets, and the U.S. has a rich fabric of wireless resellers—some of them, like MCI that uses reseller networks for its wireless services, even have national presence and national brand names. But this contradicts what seems to be a common theme in corporate strategy in the U.S. telecommunications industry: Owning your own platform is critical. At some point, when a truly competitive marketplace evolves and the law of the jungle becomes more and more the rule, a provider that relies on resale for securing a customer base could face shortages and limited access to the capacity required. In theory, facilities-based competition should also imply that interconnection agreements depend solely on commercial negotiations and that failure to reach agreements would not provoke government intervention. So, if evolution runs its course, the industry can no longer rely on lawyers to poke holes in financial data and win money in lawsuits. Instead, the competition will have to be for direct access to the customer.

The opportunities for outsourcing and reselling continue to be an important part of the discussion for the remaining case studies, but the U.S. example appears to indicate that the more successful direction for planning and strategy purposes leads away from resale and certain kinds of outsourcing and toward a single enterprise approach. Bringing the assets together, and tying them directly to the needs and interests of the customer, will be the critical challenge in a competitive market.

CONCLUSION:
SERVICE THROUGH WIRELESS PORTALS

The "best telephone people in the world" did once walk through the portals of old AT&T buildings, supporting the universal network that broke apart as competition became more and more the reality of the telecommunications sector. But the past decade has only been a prelude to the world that lies ahead in American telecommunications development.

Overcoming the divisions of local versus national authority, while simultaneously resolving the philosophical differences between the need for universal service and the goals of increased competition, will be critical to freeing up the resources, people, and institutions that will build the telecommunications networks of the future. Only wireless access can bridge the gap between those divides, but perceptions of wireless access need to change in America in order to ensure that such a goal is reached. Through innovations such as the auctions and pioneers preference programs, government can come together with corporate managers in establishing a framework for the development of the telecommunications sector so that the public policy goals of enhanced service and increased access are achieved. Only then can the best information and telecommunications services pass through open portals directly into the hands of the people who want and need them.

ENDNOTES

[1]   The Kingsbury Commitment that formed the basis of the 1934 Telecommunications Act was, in many ways, the temporary victory of universal service over competition. The government, in the words of one commentator, as the "grantor of monopoly privileges," has been "preventing or slowing new competition at every turn" since the 1934 Act so as to sustain universal service (Crandall, 1992).

[2]   There is a broad academic and judicial literature on the principles of antitrust litigation, with the titles so numerous it would not be prudent to footnote a few in fear that the broad range of opinions would not be represented. The following quote comes from "Antitrust Policy Toward Telecommunications Alliances," an address by Steven C. Sunshine, Deputy Attorney General, before the American Enterprise Institute for Public Policy Research (Washington, DC: July 7, 1994).

> Our point of departure for analysis of many telecommunications mergers is grounded in a fundamental principle: Antitrust enforcement is designed to promote innovation and efficiency. There can be little doubt that innovation, whether in the form of improved product quality and variety or of production efficiency that allows lower prices, is a powerful engine for consumer welfare. One need not look further than the AT&T divestiture to see the critical role that competition plays in spurring innovation and investment. In the early 1970's, Corning developed fiber optic cable and tried to sell this wonderful new product to AT&T. AT&T, one can surmise, probably didn't respond, "thank you very much, we'd be thrilled to adopt and rapidly deploy a new technology that will make our huge investment and undepreciated plant obsolete." Instead, it took a consortium of small long distance carriers to lay the first fiber optic network, followed by Sprint and the pin drop and MCI before AT&T laid its first such network.

Defining the Institutional Consensus: Strategic Liberalization in the U.K.

It seems that you can buy a wireless telephone and service package on any given block in downtown London these days. Advertisements for the United Kingdom's four wireless access providers are everywhere, and dozens of possible service packages have been offered to tempt consumers from all walks of life to purchase a wireless phone.

But the true nature and the extent of competition is hidden by the fanfare and flurry of advertisements and marketing. The market for wireless access in the U.K. bears what some would call the "strong illusion of competition." That strong illusion is based on a set of presuppositions about the nature of the telecommunications market in the U.K. and the potential progress for wireless access as a competitor to wireline and fixed-link services. Some of those presuppositions are similar to those in the United States, and are embodied in the predominance of the cellular model in the U.K. and the continued dominance of the main wireline carrier. Others are particular to the U.K. case and the history of the country's telecommunications infrastructure.

Even so, the goal of sustainable telecommunications development will be a critical one for the U.K., especially considering the size of the market and the relative youth of most of the service providers. Over the past decade, the number of players in the U.K. market has exploded, leading to a flurry of activity that may lead to a complete telecommunications revolution throughout the country. There is also the European connection. As part of the European Union, the U.K. has struggled to lead where other countries have been content to follow, and the marks of leadership are clearly apparent in the positioning of local service providers.

But the country's leaders are more skeptical; the hype associated with new products and services has begun to stretch the boundaries of patience as consumers wait for what has been promised by some of the established players. It is at the local level where the battle will be won or lost. Even though the U.K. does not have the size of the U.S., nor its history of local and federal division on key policy issues, local communities in the U.K. are continuing to assert themselves and their unique priorities. This tension is expressed in different ways, and the comparative characteristics of telecommunications development reflect those differences. Nevertheless, linking national regulation with local investment remains the key to sustainable telecommunications development, even in the U.K. context.

The critical piece of the puzzle will be to define a model for institutional cooperation; a model consistent with and built on the dramatic leap forward Britain has made in the past decades. This chapter, by discussing the unique characteristics of the U.K. marketplace and the role of wireless access in the future development of the U.K. telecommunications market, offers some direction in that regard. The comparisons with the U.S. and the themes raised in the previous chapter provide an opportunity to assess he direction of both of these markets, as well as the opportunities for public policy and corporate strategy to take advantage of the evolution of this particular piece of the world's telecommunications jungle. As before, the opening portion of the discussion focuses on the history of telecommunications in the U.K., then turns to the development of wireless access services, and concludes with a examination of the level and character of government policy and corporate strategy appropriate to the U.K. case.

A HISTORY OF TELECOMMUNICATIONS IN THE UNITED KINGDOM

As in the United States, the first institutions that arose to provide telecommunications services in the United Kingdom were privately owned. But, unlike the example from across the Atlantic, the early history of telecommunications development in the U.K. is marked by a slow but conscious trend by governing authorities to dominate and control the new technology.

The first telephone exchanges were constructed in the United Kingdom by the Bell and Edison companies in London in 1878 and 1879, with the patents for the technology acquired by United Telephone Company (UTC) (Hills, 1993). As the UTC began to sell exchanges in Britain and raise money for their ventures, the company ran into a number of legal and bureaucratic challenges from the Post Office. Eventually, the company was forced to assign the patents to the Post office in 1880 (Baldwin, 1938; Pitt, 1980).

But after the patents were transferred, the Treasury did not allow the Post Office to market the service. Between 1881 and 1898, the Post office installed "only 49 telephone exchanges, of which only three had more than 20 subscribers and nine had no subscribers at all" (Pitt, 1980). The critical battle was being fought for the trunk lines that provided long distance, rather than for the local exchanges. Eventually, the government decided to nationalize the long-distance lines and put them under the control of the Post Office, and the Post Office was given the mandate to compete directly with the local telephone providers.

To further fuel competition, the government granted six municipal licenses to provide telephone services, all of which, except one, failed and had to be repurchased by either the UTC or the Post Office. By the late 1890s, the National Telephone Company (NTC) had been built on the combination of the UTC, Lancashire and Cheshire Telephone Companies, but there was little economic incentive for them to invest in network construction. "Vague government intentions and a serious threat of nationalization discouraged investors from assuming the risk of funding capital expansion," as Raymond Duch (1991) pointed out. "The impending takeover had the effect of inducing complacency in the NTC directorate," according to historian Douglass Pitt (1980), "which, from 1908 onwards, had placed a moratorium on plant development." Eventually, the NTC agreed to turn over its network in 1911 after a Select Committee of the Parliament suggested nationalization.

During this period, which mirrors the interval between the expiration of the Bell patents in the United States and the establishment of the Kingsbury commitment, Britain did not experience a dramatic growth in the penetration of telephone service. The explanations for the slow penetration of the telephone before World War I range from the political economic analysis of Gerald Brock (1981)—the Post Office was trying to protect the telephone—to the sociological assessment of Douglas Pitt (Pool, 1977)—the telephone was an instrument of business in Victorian England, and not a social instrument.

Jill Hills' research suggested that there are other causes that should be considered. Hills (1977) argues that "[the] Treasury, in coalition with large users, determined tariff structures. Flat-rate tariffs and cost-based pricing resulted in cross-subsidization from small to large users and from local to long distance service." She suggests telephone service in Britain evolved out of the private wires of the telegraph service and resulted in a bias in favor of business users. As was the case for telegraph lines, subscribers had to pay the capital costs of the local loops. This is the exact opposite of the experience in the United States, where the establishment of universal service required a coalition of local communities and local service providers—a fact that resulted in the favorable cross-subsidization of residential services over business services and ensured that geographically isolated regions received telephone service. This set of factors defined the development of landline telephony in Britain, giving rise to the "top-down" investment and construction model common to many PTT administrations in Europe.

There is another critical factor, one which echoes the thoughts of Raymond Duch (1991) as presented in the first and second chapter. The inability of the government to establish an environment to protect and foster the growth of the telecommunications sector in Britain was a political failure—one rooted not in the ownership structure per se, but in bureaucratic and political constraints. Over the course of the next 60 years, those constraints continued unabated and severely hampered Britain's ability to dramatically expand its telecommunications network and keep ahead of critical technological developments.

The Period of Monopoly and Public Stewardship (1912-1984)

On January 1, 1912, Britain's telephones were effectively nationalized by the purchase of the NTC by the British government and the incorporation of the network into the Post Office's telecommunications holdings. The resulting combination was less than adequate in comparison to other countries at similar stages of economic development. As Hills outlined in her 1986 book, *Deregulating Telecoms: Competition and Control in the United States, Japan and Britain*, U.S. telephone penetration in 1906 was one phone per 15 people, while, in Britain of 1930, penetration was less than one phone per 25 people. Local telephone calls in Britain were twice as expensive as in the United States and three times as expensive as in Norway and Sweden in 1927. Waiting lists of almost a half million were not uncommon during the 1950s, and a domestic cartel controlled all equipment manufacturing and network construction through its close affiliation with the Post Office. In other words, the Post Office inherited in 1911 a run-down network demanding large-scale capital investment. Clearly, the U.K. faced a significant telecommunications gap early in the history of the sector's development.

Complaints about service and the relative poor performance of the sector prompted a number of attempts at political action. In the 1920s, members of Parliament publicly argued for the privatization of the industry. In 1932, 320 Members of Parliament supporting the national government signed a statement advocating "the adoption of a utility-type organization or the telco that would encourage a more business-oriented approach to the marketing of telephony" (Duch, 1991). These initiatives did not do much to change the operation of the Post Office, leading to increased frustration among certain segments of the British political leadership. In 1969, the Post Office Act changed the status of the Post Office from a government department to a public corporation that still retained control of posts and telecommunications. Nevertheless, the change did little to improve the ability of the government to overcome three specific constraints: finances, employment, and corporate strategy.

The Post Office could not borrow privately and had to rely on government expenditure for its investment. This policy hamstrung its ability to manage its resources in a way that would improve the progress of the sector. As has been the case for almost every PTO organization throughout the world, the finance constraint also consisted of political decisions to keep prices rising at a rate less than the rate of inflation, causing gross margins to shrink and forcing the company to put off infrastructure investments that would have helped to improve services.

The labor unions also restricted the range of possible institutional transformations, and its inflated workforce in 1982 included 50% more employees than the French PTO. This difficulty was compounded by the fact that increases in real staff costs for the telecommunications sector outstripped increases in costs for other sectors of the economy by a substantial rate.

Finally, British Telecom was compelled by government policy to choose British manufacturers, and all costs for the development of new digital exchanges were jointly shared by British Telecom, Plessey, and GEC. This additional cost depleted the resources of the company, but not as much as the premium that British Telecom had to pay for the equipment relative to the cost of other potential suppliers from other countries. In sum, the constraints placed on the company by the political institutions undermined the ability of the company to invest in its network, its people, and the services it provided to its customers.

The Preparations for Privatization

In 1977, Charles Carter was commissioned to evaluate the performance and organization of the new corporation, and his assessment was not favorable:

> The report was highly critical of the Post Office's management structure, financial management and its policies towards the introduction of modern technology. Carter found that the Post Office, as a result of the 1969 Act, believed itself to have thrown off the shackles of Government accounting and to be operating as a commercial organization. However, its attitudes toward costing, toward price decisions and towards its relations with customers were far from those that should arise in a true commercial operation. (Garrison, 1988, p. 4)

This negative assessment of the Post Office's ability to administer the telecommunications network in the face of increasing economic and technological pressure came just before Britain elected the Conservative party to power in 1979. As Margaret Thatcher pointed out after her election, "If we stand idly while we watch others become more efficient, the time will very quickly come when what happened to British typewriters, and British cameras, and British motorcycles will also happen to the British information technology industry" (Garrison, 1988). To show her commitment to the reform of the sector, one of her early moves was to appoint Kenneth Baker as minister for information technology and placed the extension of competition policy to the telecommunications sector high on the agenda.

Baker and the Conservatives immediate set the groundwork for the Thatcher administration's first major policy measure in this area: the 1981 Telecommunications Act. The Act opened the market for telecommunications equipment, manufacturing was opened up to competition, and licenses to provide private value added network services (VANS) were granted to operators who met certain technical specifications. Parliament also approved the creation of British Telecom, which became the new public corporation to run the national telephone network, thereby separating the Post Office from the telecommunications business. Finally, Parliament decided to privatize and sell 100% of Cable & Wireless (The Marconi Company's grandchild) in 1983.

The first major company to be licensed as a value-added service provider was Mercury in February, 1982. The license, granted and overseen by the Department of Trade and Industry (DTI), allowed Mercury to run a digital network that provided services directly to business users. A general license was issued soon after, allowing other competitors into the same market. In addition, Parliament empowered DTI to license two cellular service providers in 1982, with one license eventually going to Mercury and the second to BT.

The Telecommunications Act of 1984
and the Establishment of Oftel

The Conservatives were not satisfied with the liberalization of the market, even though many of the steps they took during this 5-year period were revolutionary in comparison to the highly regulated and tightly controlled public telecommunications operators in Europe and in other countries throughout the world. The Government took the final step and chose to privatize BT by selling off all of its shares in the company over the course of the following decade.[1] The Act further defined the regulatory authority of government agencies and the possible competitive roles of various private-sector players.

Perhaps the most significant element of the 1984 Telecommunications Act was the establishment of a new regulatory authority: Oftel, which is the abbreviation for the "Office of Telecommunications." Oftel was housed in the Department of Trade and Industry (DTI), a government portfolio with a broad authority over the industrial policy and economic development strategy of the British government.

The relationship to the Department of Trade and Industry is perhaps unique from the perspective of comparative telecommunications studies. The Ministry is formally the ultimate decision maker, but the analysis of the regulatory issues and the decision making in practice occurs at Oftel (Tyler and Bednarczyk, 1992). In that regard, almost everyone considers Oftel to be a comparatively less intrusive authority when compared to the American FCC. Oftel's authority is also constrained by the other institutions with which it shares regulatory responsibility. The Radiocommunications Agency is responsible for radio spectrum management and the Monopolies and Mergers Commission (MMC) plays a central role in determining the appropriate regulatory policy for the sector.

When it comes to implementing competition policy, this structure and division of responsibility has a number of advantages. For example:

> The mechanism for changes in price regulation for British Telecom is the adoption of amendments to BT's license that are negotiated between Oftel and BT. If BT fails to agree to Oftel's proposal, Oftel itself cannot impose its view, but can refer the matter to the Monopolies and Mergers Commission (MMC) which has far reaching investigatory and enforcement powers. In practice, the threat of doing so has invariably been sufficient to ensure agreement. (Tyler and Bednarczyk, 1992, p. 19).

Licenses for telecommunications providers follow the pattern of other common law in the British legal system; licenses are evolving documents, and are not fixed in time as they are in the United States' system of jurisprudence. Alterations in regulation and legislation become part of the license of a telecommunications provider. Licenses, therefore, form the critical ground for the ongoing public debate about telecommunications service in the United Kingdom. The construction and alteration of those licenses provides, in many ways, the most telling historical signposts for the discussion of Britain's telecommunications sector because they define and codify changes in corporate institutional behavior.

The first such instance of Oftel referring a license modification to the MMC occurred in 1988:

> In July 1988, OFTEL made its first reference to the MMC, asking it to investigate chatline and message services. The Monopolies and Mergers Commission's report on these services was published in February 1989. It concluded that the provision by BT of chatline and message services by means of its public switched telephone networks, and the provision by BT of a telecommunications service to other persons enabling them to provide chatline and message services by means of its public switched telephone network, operated or might be expected to operate against the public interest in that, due to the case of access to the services and the terms of the contract between BT and its customers, the customer had inadequate control over the types of service which could be accessed and over the costs or charges that might be incurred for the issue of the services, which significantly impaired the value and quality of the telephone service to the customer. (OECD, 1992, p. 94)

The result of the decision was that the BT license was modified so that the company had to provide its customers, in advance, the transparent costs for chatline services.

This combination of institutional linkages and bifurcated responsibilities does place Oftel in a comparatively weaker position than other regulatory institutions throughout the world, but that has not prevented Oftel from taking a very active role on certain key issues. On issues of customer satisfaction and responsibility, for example, Oftel has become a critical watchdog on industry matters, often earning the ire of the carriers (particularly BT, which, as the dominant provider, usually is the main target of Oftel inquiries). In addition, Oftel has instituted some direct controls, requiring that BT increase its optical fiber plant from 2.045 million kilometers in March, 1992 to 3.53 million kilometers by the end of 1997. So the "light touch" is not often as light as would initially appear—and perhaps that touch will continue to become heavier as the relatively young agency continues to tread in new areas of regulation and sector reform.

But back in 1984, the newly created Office of Telecommunications had a long way to go before establishing the presence and authority apparent in the preceding passages. Its first critical task was to ensure the implementation of the most important provisions of the 1984 Telecommunications Act: the licensing of new competitors and reconstitution of licenses for existing market participants, most critically British Telecom. As the privatization of BT moved forward and as Oftel began its work to increase the level of competition in the telecommunications industry, a framework for constrained competition developed. This framework, defined largely by the structure of licensing arrangements in the sector, drove the institutional behavior of market participants throughout the remainder of the 1980s.

1984 to 1997: The Emergence of Constrained Competition

Following the Telecommunications Act of 1984, there are three broad categories of licenses in Britain:

- PTO (Public Telecommunications Operators) licenses, which generally provide for the provision of infrastructure and strict service and competition obligations;

- Non-PTO licenses for local, regional services and interconnection with PTOs; and,

- Class licenses, generally to permit operation and connection of terminals, classed as telecommunications systems under the Act.

Because these licenses define the strategies of corporate participants, it is not surprising that the framework of competition that followed from the passage of the 1984 Telecommunications Act would reflect the classes of licenses outlined. Competition between providers occurred within each of these classes, but was limited by the nature of the licenses granted and the levels and nature of the investments made by the industry participants. In addition, competition between providers with classes of licenses had yet to take shape. This was a period of constrained competition, a time of preparation for what was perceived to be a broader competitive environment in the 1990s.

The most significant licensing change induced by the 1984 Telecommunications Act was the alteration of the BT license. No longer a government owned and operated network, BT was positioned as the dominant carrier in a market that was to have at least one major competitor: Mercury Communications Ltd. Section 8 of the 1984 Telecommunications Act places PTOs under an obligation to provide services or connection on a nondiscriminatory basis, as defined by the license. In the first category of licenses, that of the PTO operators, a duopoly policy was put into effect; Mercury Communications was licensed as a PTO to provide direct competition to BT for all telecommunications services.

In the beginning, Mercury was an independent operator with little of the infrastructure required to compete directly with BT for market share. Even after the purchase of Mercury by Cable & Wireless,[2] the amount of investment required to build a competing network would be in short supply for the company, forcing it to develop a strategy common to many carriers in the early stages of competition.

Through the intervention of Oftel, Mercury would use BT facilities to directly compete with the company for market share.

That strategy required a clear definition of the costs and character of interconnection, a factor that remains a common problem for competition in British telecommunications policy today. Working out the interconnection policy took more than three years, and focused on the difficult political question of interconnection between the two PTOs. In late 1982, a basic agreement between the two companies was reached, but it was not until October, 1985 that Oftel published a determination on the terms of the full interconnection agreement at the switch level between the two networks (Gillick, 1991).

When Mercury was licensed, the government promised to license no further fixed link national PTOs to compete with the company or with BT for at least 7 years. During that period of time, Mercury began to make inroads into certain portions of BT's market share, but did not erode BT's position as the dominant PTO in Britain. At the end of 1993, BT still held about 89% of the market for voice telephony (Lehman Brothers, 1993).

Another major investment in wireline capacity has occurred since the Telecommunications Act of 1984: cable television. In 1983, the government published a white paper entitled "The Development of Cable Systems and Services," which set out the policy for rolling out cable systems throughout the country. In combination with the Telecommunications Act of 1984, there was great expectation for cable to penetrate the British market quickly through the second half of the 1980s (Smith, from Dutton, Blumler and Kramer, 1987). Those expectations fell flat, though. The initial level of investment was low, and throughout the 1980s, there was much discussion on why the cable market in the United States was exploding while little or no penetration occurred in the major urban centers of Britain.

Perhaps the critical government action to spur the development of the cable industry was the raising of the ban in July, 1989, on significant U.S. investment in the cable industry. The immediate result was a tremendous influx of investment from major service providers and equipment producers such as NYNEX, Comcast, and U.S. West. With regard to the services that cable operators could provide, the government introduced one critical regulatory innovation in the early 1980s that no other country in the world attempted to implement. Cable operators were allowed to provide telephony in conjunction with cable services, first in association with one of the two carriers (BT or Mercury) and then, after 1990, on its own platform. This was the clearest attempt to provide direct competition between two classes of licenses: the PTO and non-PTO industry participants. At the end of 1994, there were 106 operators authorized to provide telephony in their franchise areas, all of them potential competitors to (or partners with) Mercury and BT. But this growth would not have been possible without the finalization of the duopoly review, which marks the beginning of the next important phase in the development of the U.K. telecommunications sector.

The Duopoly Review: Goals and Outcomes

Little direct competition in residential telephony had emerged in the U.K. market from the advent of the 1984 Telecommunications Act to the beginning of the 1990s. As one observer put it, the market for services resembled more of an "oligopoly in which BT and Mercury—or, more accurately, C&W—are the dominant players, followed by Racal, British Aerospace Communications, the Regional Bell Operating Companies (RBOCs), North American cable companies and so on" (Gillick, 1991). Each of the non-PTO players had been given its particular geographic and/or service niche, and there was little effort made (and little opportunity given) to cross over into other niches and directly compete for services. BT was still clearly the dominant carrier, and Mercury had focused its attentions on businesses and intensive users of telecommunications services rather than the mass market.

Oftel responded to these market conditions as part of its Duopoly Review, a process undertaken through 1990 and 1991. The thinking of Oftel's Director General during the Duopoly Review, Sir Bryan Carsberg, reflected the evident tension between adherence to a strong competition policy and the difficulties inherent in inducing competition in the telecommunications sector. The following quote comes from a statement issued by the Director General on March 5, 1991, entitled "Competition and Choice."

> As existing competitors extend the scope of their operations and more competitors come into the market place, losses of economies of scale continue to mount and a time will come when the benefits of additional competition cease to outweigh the costs. The normal disciplines of the marketplace can serve to guard against the ensuing danger that higher costs are passed on to customers. Policy should be to stop assisting entry by regulatory means while continuing to encourage entry and use all regulatory weapons to prevent its being impeded by anti-competitive practices. New competitors will then succeed if and only if they can win their success through performance—by superior efficiency, superior quality of service or innovative services. (Oftel, 1991, p. 3)

This statement reflected the prevailing view within Oftel that the regulatory assistance granted to Mercury was no longer necessary and that it was possible for the two companies to compete on equal terms through established interconnection agreements. The result of this consensus within Oftel and the government was a set of proposals enacted after the close of the Duopoly Review proceedings, with some of the most significant including:

- Ending the duopoly policy at the local, trunk and international levels;

- Removing the regulations which required cable TV operators to offer telephony only in conjunction with BT and Mercury;

- Permitting companies with partially common ownership with a fixed telecommunications operator to tender for new franchises and provide and carry entertainment services;

- Permitting the assets of utilities, such as electricity, rail, and waterways, to be used for competing in the telecommunications market; and

- Allowing mobile operators to provide fixed communications services. (Pye, Heath, Spring, and Yeomans, 1991)

These policy changes were focused on the diversification of the telecommunications market in the U.K., and, in many ways, are more progressive and expansive even than those enacted in the U.S. with the passage of the 1996 Telecommunications Act, if only because the doors were opened wider to companies like electric utilities and rail companies. The implementation pattern that followed was consistent with this objective; in the 3 years following the conclusion of the duopoly review, Oftel granted more than 60 PTO licenses for the provision of local or national telephone service.

But, even with this expansion in the number of players in the telecommunications marketplace, BT has been able to retain a large portion of its market strength, although in each year subsequent to the duopoly review, public comments that BT would lose a high percentage of market share were common—even from BT senior executives. But no other company or combination of competitors has dramatically eroded BT's market share. That is because BT has done a tremendous job in protecting its market share, both in terms of the business decisions it has made and the active role it has taken in shaping the public policy process. In that regard, any discussion of the impact of the duopoly review should focus largely on its impact on BT and the company's position vis-à-vis other competing companies.

The Impact of the Duopoly Review on BT's Market Position

As the dominant provider of all telecommunications services in the U.K., BT has had to face competition from both the private sector and the regulatory audiences. In that regard, the challenge of functioning in the U.K. marketplace has been similar for BT and some of the Regional Bell Operating Companies in the U.S.—because the very nature of telecommunications service provision has been tied to the expectations of regulatory audiences, the structure of the company and the services offered have been a reflection of both market pull and regulatory dictate.

The character of "competition with the regulator," in this case Oftel and other institutions like the MMC and the DTI, has been rather fierce since the conclusion of the duopoly review. Traditional constraints on the company have persisted, embodied in license provisions which have been actively enforced by Oftel. And some of those provisions sound very similar to those imposed on the RBOCs, such as Section 3 of the license which requires BT "to promote the interests of customers, purchasers and other users in the United Kingdom (including, in particular, those who are disabled or of pensionable age) in respect of the variety of telecommunications services provided."

As has traditionally been the case for the regulation of the dominant carrier in an environment of emerging competition, the language of the license attempts to define areas where cross-subsidization of service would be a detriment to competition.[3] In addition, BT was placed under an obligation to permit other authorized systems to connect to its system, with the quality sufficient enough to ensure that any customer of one public telecommunications system should be able to make calls to any customer of any other system.

BT has succeeded, so far, in achieving what Ian Ash, BT's Director of Marketing, called in 1992 the company's "primary priority," namely to defend its position as the preeminent telecommunications provider in Britain. BT has been able to focus its resources so as to reduce the impact of competition in certain critical areas during the 1980s, such as leased lines and circuits. Since BT's privatization in 1984, Iain Valliance estimated in a 1993 speech that BT had spent more than 20 billion pounds building "what is arguably Europe's most modern network." At the same time, it has moved aggressively in its own market and throughout the world to establish itself as a premier telecommunications service provider, first by planning its own global network, then by purchasing a substantial stake and eventually offering to purchase U.S. long-distance provider, MCI.

This combination of investments in reaction to perceived opportunities to counter regulatory constraints has given BT an edge in keeping ahead of the second area of competition—other telecommunications companies. There appears to be little possibility of Mercury significantly eroding BT's present market share. Beyond the city of London, Mercury's impact has been "patchy," according to Andrew Adonis of the Financial Times (1995). As of 1994, Mercury had only 780,000 residential customers to match BT's 20 million. The lack of performance in the wake of the duopoly review sparked a management row which continued through much of 1995 and 1996, causing the ouster of many of the company's senior executives. To compound the problem, as the following table shows, Mercury also receives comparatively low marks on efficiency and service in the critical business telephony market. BT is reporting considerable increases in the SDN and other higher bandwidth wireline services. With BT bringing prices down and girding itself for further competition, the reviews of Mercury's potential for making serious inroads into the local market were mixed.

In late 1996, Mercury's parent company, Cable & Wireless, took steps to reverse that trend. Cable & Wireless acted as a catalyst for the consolidation of the cable industry in the U.K. by drawing together NYNEX CableComms, Bell Cablemedia and Videotron to create a new competitor in the local exchange. But even this move to vertical integration and facilities-based, wireline competition may not be enough to erode BT's market share—over the last three years, the direct impact of the development of cable television capabilities on competition for basic telephone service has been limited.

Some have described the construction of Britain's cable system as the precursor to much more significant levels of competition. Five years after the ban on foreign investment in cable franchises was lifted, the franchise map of England was almost complete, with 127 broadband franchises covering roughly 14 million homes, with American companies owning roughly 75% of the business (Oftel, 1995). But it is the growth of cable telephony that really captured the imagination of regulators and investors alike, especially because of its potential implications for markets in the United States. After an initial rush of optimism, and reports of almost overnight penetration rates of 25% among households in franchises with cable television services, estimates were that 103,000 businesses and 700,000 residences will be connected to cable telephony by the year 2000, generating 38% of cable's revenues (Oftel, 1994).

Other potential competitors, such as energy companies and railways, have only begun to organize themselves to play a role in the market. Energsis was established by the National Grid Company (NGC) to compete in the long-distance services market. The NGC, recently created as part of the U.K.'s continuing experiment in the deregulation of the electricity and power industry, has leveraged its telecommunications assets in an attempt to compete with BT and Mercury, largely in the business services market. But the losses have been immense—more than 52 million pounds in 1994 and 1995. The NGC has been looking to unload its investment, even while the National Rail Company looks for investors to take control of its future telecommunications offerings.

Competition by local cable operators and other telecommunications companies for residential telephony services is considered by many commentators to be the cause of recent reports from BT that growth in residential telephony lines for the company have come to a halt, and have actually begun to decline in recent quarters. The first reported decline occurred in the second quarter of 1995 and, in the third quarter, BT reported adding 120,000 new lines, but suffered 150,000 disconnections. Nevertheless, this still represents a small fraction of the total number of subscribers for voice telephone service presently held by BT—and considering the high rates of churn in the cable telephony markets, it is likely that BT's market share in residential markets will not disappear anytime soon. If AT&T's experience in the long distance market holds any precedent, BT will face a loss of between 20% to 30% of its market share and may then find a stable level, as it has in the competition with Mercury for the business market.

In addition, BT continues to face difficulties in improving its efficiencies relative to other global telecommunications providers. Even though BT's downsizing has continued through the 1990s, with the company shrinking another 11% during 1995, lines per employee comparisons still show BT stuck in the middle, between 7th and 9th position over the past decade (Cracknell, 1996).

So things are not all strawberries and cream for the U.K.'s dominant telecommunications provider. From BT's perspective, the U.K's regime for telecommunications legislation has advanced competition for landline telecommunications services. Domestically, the result has been decreased prices, but not the substantial weakening of BT's market position. In turn, BT has been able to invest internationally and position itself in potentially important markets on both sides of the Atlantic.

In that regard, the overall result of the last decade of investment and change since the enactment of the 1984 Telecommunications Act has been the emergence of constrained competition. With BT in a strong position to retain market share for the foreseeable future, it will be critical to establish a new model for institutional interaction that removes the constraints on other marketplace participants so that they can continue to grow in strength and viability in the face of BT's continued dominance. If not, Britain could find the development of the telecommunications sector stunted by the same kinds of political failures that undermined the development of its networks in the early part of the century.

The European Context

For the U.K., nonetheless, the domestic market is far from being the only focus for the country's telecommunications providers. Britain's conscious policy has been to position itself as the leader in the European Union with regard to the telecommunications sector.[4] It also has played a central role in the development and promulgation of the European Commission's present policy advocating increased competition in the provision of telecommunications service and infrastructure. For evident reasons of comparative advantage and commercial interest, this opportunity to compete throughout Europe has been a significant part of the institutional consensus supporting the continued liberalization of the telecommunications regime to date.

While Britain was rewriting telecommunications history in 1984, Europe began down the long road to market liberalization. Beginning with the Green Paper in 1987 and continuing throughout the 1980s and 1990s, the European Commission and individual countries pushed the agenda of sector liberalization. After significant legal challenges, the Commission mandated the deregulation of the marketplace through the liberalization of the telecommunications market in all member states. The results of an extensive consultation procedure lead to a document, published in 1993, which stated that the deregulation of the market would occur by 1998. The Commission also commissioned and published what is known as the Bangemann Report in early 1994, further making the case for competitive changes within member countries.

Ironically, Britain can safely be said to be far ahead of the curve when compared to its European compatriots—liberalization and privatization have given the British an advantage. For example, value-added networks have generated almost one billion dollars in revenue for U.K. telecommunications companies during the 1980s, as opposed to $666 million in France and $428 million in Germany (*Communications Week International,* June 4, 1990). The depth of experience for companies like BT, Mercury and the cable television companies should not be underestimated—and the fact that further liberalization is being resisted by the politicians of France and Germany may be a tacit recognition of British strength in this area. The irony is that Britain has generally been the advocate of national-over-European determination on broader issues of sovereignty, but that the implementation of the Commission's directives offers a competitive advantage to British service providers.

Why has the Britain been able to chart a dramatically different course than most of the other countries of Europe? Duch (1991) suggests that the "British policy response is different because of the decidedly pluralist nature of its policy-making institutions." He points to three important ramifications of pluralism:

- Interest groups are much less beholden to the state and more comfortable with the lobbying of civil servants, ministers and members of Parliament;

- Barriers to participation in the policy process are much lower in the United Kingdom; and,

- Pluralist institutions in Britain permit policy outcomes that are build on the support of minimum winning coalitions.

Duch seems to feel that the British model will win out as the European Union charts its course towards consolidation. As he puts it:

A corporatist model could be imposed on those European institutions where European interests have very centralized organizations and are willing to participate in consensual decision-making forums. Similarly, the statist model could only succeed if the European Commission were able to command the degree of authority that is exercised by the French state, for example. Neither of these situations is probable because of the important role of the United Kingdom in the European Community. The country's pluralist structures (antagonisms between labor and industry and the passive role of the state bureaucracy) are inconsistent with corporatist or statist approaches. On the other hand, because pluralism accommodates diversity by its very nature, this institutional model is most likely to be adopted. (Duch, 1991, p. 270)

If Duch is right, and it is the ability of the British system to bring together diverse interests on common ground through an open political process, then the competition policy articulated in the early 1980s must be seen as the source of strength of the present British position. It is a policy that has benefited all of the major institutions and players significantly over the past decade and puts each institution in the position to benefit further in the years to come.

In many regards, the U.K. has achieved a consensus in areas where the U.S. has failed. The continuing legacy of local versus state conflict, although a reality in the U.K., is certainly not as pronounced as it is across the Atlantic. The advantages of a smaller market and a unitary state show in the telecommunications policy of the U.K.—even though the Americans may counter, in turn, that the pattern of apparent recklessness that launched Silicon Valley and other great examples of U.S. economic expansion seem to indicate that chaos, rather than consensus, is the best course for economic development.

The ability to reconcile diverse interests, within the context of policymaking in the unitary state, has facilitated the implementation of a new philosophy of governance in the U.K.: competition and privatization of the public sector. No matter what the final judgment of history is for Margaret Thatcher's reign as Prime Minister, it can not be denied that the privatization of British Telecom, among other formerly public properties, brought billions of dollars of investment into the British economy and dramatically changed the nature of the economy and society. The acceptance of this model for political and economic relations is seen in the Conservatives' continuing hold on power and the left's repositioning toward the center on issues of competition policy and public sector involvement in economic affairs.

The diminishment of the left in the U.K.'s telecommunications policy debate echoes the rise of the advocates of strong liberalization and the more radical techno-libertarian approach around the world. A space has been opened up for experimentation in the telecommunications sector, and that space is likely to expand as long as the apparent successes of increased investment and service penetration draw the attention of politicians and corporate executives.

The transformation of the corporate institution is also part of the legacy of privatization, deregulation, and change in U.K. telecommunications sector. For BT, it has meant increased profitability and the opportunity to access telecommunications markets in Europe and throughout the globe. Without the liberalization of the U.K. market, it would have been that much tougher for BT to get a substantial head start over other companies in the developed world, and even take active roles in competing in a market as advanced as the United States. Britain's ability to mobilize expertise and capital in the European market can only help BT as it continues to expand beyond its previous role as a state-run telecommunications monopoly.

For companies like Mercury and other potential direct competitors with BT (such as cable telephony providers), it has meant an opportunity to compete and invest in a growing market. The expansion of the market has made the game of telecommunications more than zero-sum; with increasing subscriber bases and the future of enhanced services, getting a slice of the pie and the opportunity to get more of the pie has been incentive enough for many investors.

But that consensus can only go so far. As the U.S. example shows, even with the implementation of the 1996 Telecommunications Act, the subsequent onslaught of interconnection agreements, and creation of new telecommunications service providers, the coherence so sought after by politicians looking to explain policy to constituencies is virtually impossible in the world of bytes, bits, standards, transmission technologies, and switch level politics. Considering the lack of competition at the local level and the likelihood that the liberalization in other markets will soon close the gap, the U.K. has much to do if it is to keep its lead in regulation and service provision.

Wireless has begun to make its own mark on local competition in the U.K., and this is the next area for discussion. What is the specific history of wireless communications in the U.K., and what are the possibilities for using wireless access as a lever for increasing the pace and impact of liberalization?

WIRELESS COMMUNICATIONS IN THE U.K.

The U.K. is the origin point for the modern story of wireless communications, in great part because of the efforts of one man: Gugliemo Marconi, the inventor of the wireless telegraph and telephone. Although born in Italy and closely tied to his home country, the bulk of Marconi's work was done in England, and Marconi applied for the world's first patent in wireless telegraphy on June 2, 1896 (British Patent No. 12,039). The bulk of initial investment came from Britain's private sector, and, perhaps most significantly, the first contract for Marconi came from the British Government.

The first major application of wireless communications in Britain involved the construction of a global communications network for the British fleet in the early years of this century.[5] The increased international tensions and the rise of a more powerful Germany brought Marconi's invention to the forefront of naval planning as both Britain and Germany raced to put in place a radio communications network for their military forces.

But commercial applications of wireless communications were put in abeyance by the same government: The consolidation of the government's control over wireline communications (both telegraph and telephone) created great resistance within the government to the development of wireless telegraphy or wireless telephony.[6] Even though Marconi had developed the first wireless telephone by 1908, the Wireless Telegraphy Act of 1904 had constrained his ability to operate to such a degree that the limited efforts of his company to create a commercial service were to no avail. More than 50 years after Marconi developed a wireless telephone, BT (then the Post Office) first offered a radiophone service, relying on 55 channels in the VHF band to carry voice traffic for commercial users.

Ironically enough, the government that discouraged the development of wireless communications for fear of competition at the beginning of the century is now advocating the further development of wireless telephony at the end of the century to enhance competition. Establishing direct competition between wireless and wireline communications is at the forefront of Oftel's stated policy:

> I believe that other European nations will come increasingly to recognize the desirability of competition in operating networks. I also believe that they will find a certain inevitability in progress towards competition. Several countries have introduced competition in mobile networks and once mobile communications becomes a comprehensive competitor with fixed communications, as I believe it soon will, competition for basic services will have arrived, as it were, through the back door. I also believe that com-

petition in value added services and in data services will lead to the acceptance of competition in basic voice services: The distinction between basic services and value added services will become harder to draw and will cease to have a clear significance as the concept of basic service changes under the impact of developing technology. (Oftel, 1991)

This speech, given by Oftel General Director Sir Bryan Carlsberg at the tail end of the duopoly review in 1991, offers a broad-reaching perspective on how wireless communications could impact the marketplace. Wireless competes directly with wireline services for basic voice telephony. Then, value-added and basic services converge on wireless and wireline networks, creating a broad framework for competition throughout the telecommunications industry. This vision is embodied in the "Phones on the Move" report discussed earlier. In setting the foundation for the establishment of Personal Communications Networks in the U.K., Phones on the Move was born of the same conviction that wireless and wireline networks would and should directly compete.

There is good reason for the regulators at Oftel to be optimistic about the prospects for direct competition in Britain between wireless and wireline communications systems. The dramatic growth of the subscriber base within the country bears out one simple fact: More and more people are turning to wireless communication systems for their personal and professional communications needs.

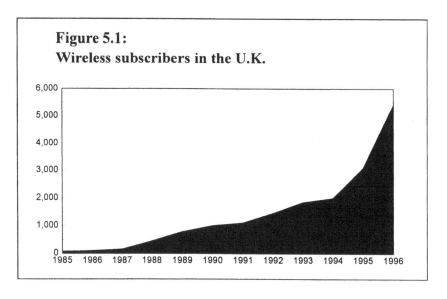

Figure 5.1:
Wireless subscribers in the U.K.

Note: Adapted from public sources, Oftel (1996) and Donaldson, Luftkin and Jenrette (1996)

This growth has been driven by the four major players in the market for wireless access today: Vodaphone, Cellnet, Mercury One-2-One, and Orange (Hutchinson Telecom). Vodaphone and Cellnet were licensed as cellular providers in the early 1980s, while One-2-One and Orange are both licensed as Personal Communications Networks (PCN). The total increase in subscribers, as well as the breakdown of the subscriber base for each of these four companies, as of the beginning of 1996, is shown in both figure 5.1 and table 5.1.

The four wireless operators have all benefited greatly from the increases in the subscriber bases. Other wireless communications services have grown almost as dramatically. The subscriber base for pagers, for example, has expanded to almost 800,000 since 1980. But the flurry of growth and activity in the sector has not yet set a firm foundation for direct competition between wireline and wireless access, in great part because of the institutional constraints placed on the development of the sector. In order to ensure a vibrantly competitive market, Britain will need to achieve a new institutional consensus to push the penetration of wireless communications systems.

Table 5.1:
Subscriber Growth for UK Providers

Operator	Subscribers (4/96)	Subscribers (4/95)	Percentage Growth
Cellnet (analog)	2,036,000	1,699,000	19.84%
Cellnet (digital)	353,000	34,000	938.24%
Vodaphone (analog)	1,930,000	1,645,000	17.33%
Vodaphone (digital)	520,000	173,000	200.58%
Mercury	410,000	260,000	57.69%
Orange	488,000	145,000	236.55%

Note: Adapted from public sources, Oftel (1996) and Donaldson, Lufkin and Jenrette (1996)

The next section discusses the institutional framework for the development of wireless communications and telephony in Britain. It begins with a discussion of the initial cellular duopoly and its impact on overall sector development in the 1980s, then turns to analyze a failed experiment in local wireless access, CT-2 (Telepoint), identifying the structural and institutional constraints that lead to its demise. Finally, the discussion focuses on the continuing experiment: Personal Communications Networks (PCN). The end of the section outlines some of the institutional barriers that need to be overcome in the redefinition of wireless access, setting the stage for a discussion of how the goals of strategic liberalization can be used to further spur the development of Britain's telecommunications sector.

The Initial Duopoly

As Britain began to consider the opportunities for the deregulation and liberalization of the telecommunications sector, worldwide recognition of the commercial potential market for wireless communications systems began to attract the attention of regulators and industry participants. In the early 1980s, at just about the same time as the U.S. considered its options for licensing the first cellular networks, the Department of Trade and Industry opened the discussion of possible applications for nationwide cellular telephony licenses.

The government invited applications for two 25-year licenses in 1983 and, of six applicants, two were chosen as the basis for a cellular duopoly: Vodaphone and Cellnet. Initially, Vodaphone was owned by Racal (80%), Millicom (15%), and Hambros Bank (5%). At the end of 1986, Racal Millicom became a 100%-owned subsidiary of Racal, and then was floated as its own limited company. The other company licensed to provide cellular services, Cellnet, was a partnership between British Telecom and Securitor; in 1991, BT increased its ownership to 60% and Securitor retained 40%. The two competing mobile communications networks began service in the United Kingdom in January, 1985 and remained the sole providers of wireless telephony until the licensing of CT-2 (Telepoint) systems later in the 1980s.

The Common Interface: TACS, GSM and DSC-1800

The Total Access Communications System (TACS) was the institutional response of a government that wanted to position Britain on the cutting edge of equipment manufacturing throughout Europe. The U.K. had already fallen behind the United States, which had its own specifications for analog mobile radio (AMPS) and the Nordic countries (NMT-450 and 900); the thought that Britain could tie the development of cellular systems locally into a springboard for international equipment sales led to the development of system standards.

The U.K. borrowed from the AMPS standard to construct TACS. Companies modified the AMPS specifications to meet the European mobile radio frequency allocation (862 to 960 MHz) and the European channel spacing practice. Over time, the E-TACS standard was developed to improve upon the original system and apply it to a new frequency band just below the initial applications for TACS.

Even with its initial difficulties, TACS has been rated highly as the standard backbone for providing cellular service in Britain (Balston and Macario, 1992). It has also grown with the times. One of the provisions of the licenses granted to Cellnet and Vodaphone was that the carriers cover 90% of the U.K. population by 1990, a goal achieved by both companies in mid-1987. The two TACS systems still in operation today handle more than 3.2 million subscribers, indicating TACS has been up to the task of managing Britain's cellular boom. From the viewpoint of competition policy and international trade, TACS has also been successful— the standard has been exported to a number of countries for use in the construction of analog cellular systems, with more than 1.5 million subscribers throughout the world utilizing TACS systems at the end of 1993. TACS has made limited inroads on the European continent. U.K. equipment providers still face competition from other companies, especially Noika and Ericsson, both of which actively pushed the NMT standards developed earlier in the decade for mobile telephone systems throughout Scandinavia.

The result of this competition between equipment providers and technical standards was the fragmentation of the market for cellular radio equipment in Europe. That factor was certainly on the mind of the European Commission in the mid-1980s. The advent of digital communications, and the rise of the computerization of telecommunications networks, made it clear that various digital systems for the provision of wireless communications systems would soon be rolled out. A number of working groups, under the sponsorship of the Commission, had begun to test various digital systems, with the goal of developing a "Global System for Mobile Communications," or GSM (Balston and Macario, 1992).

A number of digital transmission techniques were suggested for the basis of the GSM standard, the most significant of which (TDMA, CDMA, and FDMA) have already been described in the third chapter. The choice of TDMA as the transmission standard in 1987 set the foundation for the first articulation of the standard by the Commission on European Post and Telecommunications (CEPT). In 1989, the responsibility for the specification of the GSM standard passed to the newly formed European Telecommunications Standards Institute (ETSI).

The difficulty facing the U.K. during this period was the intensive investment made in TACS and the lack of any digital transmission system being advocated by U.K. equipment producers that could meet the requirements for high mobility. The government faced the prospect of falling behind again to other European equipment manufacturers with experience in producing the components for the future GSM system.

The U.K. responded by establishing a comparative advantage in service provision. While other countries thought of GSM as the standard to facilitate the transition from analog to digital transmission in the cellular bands, Britain began developing a framework for licensing Personal Communications Networks (PCN) in the 1800 to 2000 MHz band. The goal of PCN would be the establishment of wireless access services at a lower cost than the traditional cellular services, with the overall objective of increasing competition in wireless services.

Instead of relying on U.K. manufacturers to develop a standard for the country and face the prospect of falling behind yet again, the Department of Trade and Industry used the European developments in this area to their advantage:

> The DTI recognized from the outset the importance of adopting internationally agreed upon standards for the ultimate success of PCN, including the conditions for roaming between PCN networks. It was stipulated that the specification must be based either on the emerging digital European cordless telephony (DECT) standard for cordless private branch exchange (PBX) or on the pan-European GSM digital cellular mobile system.

> All PCN licensees stated their preference for developing a standard for PCN based on GSM, redefined for operation in the 1710 to 1880 MHz band, incorporating initial PCN requirements. This view was consistent with one of the key recommendations made by the ETSI Strategic Review Committee on Mobile Communications in Recommendation 8—Digital Cellular System at 1800 MHz (DCS 1800): "GSM must be asked to elaborate an enhancement to the GSM standard for using frequency bands compatible with the plans of CEPT and located in the vicinity of 1.8 GHz. This new version of the standard should be aimed primarily at providing a service for handheld or pocket terminals in densely populated zones and must be suitable for pan-European implementation." (Hadden and Knight, from Balston and Macario, 1992, p. 224)

Even though the U.K. had fallen behind in terms of equipment production and development, they were moving fast ahead on service implementation, pushing the European standards-setting bodies and putting solutions into place before other European countries. In turn, a number of British companies acquired direct experience in producing components of the system and the support infrastructure for DCS 1800—and the U.K. has become the major testing ground of the new technologies of all the European equipment manufacturers. With an eye to serving the critical European market over the long term, the U.K. established a comparative advantage by licensing providers, and turned a weakness in equipment production into a competitive strength.

Retail Competition and the Illusion of Choice

The method by which comparative advantage has been established, though, threatens to hamper the continued development of the sector. This is due to the complicated arrangement between retail service providers, wireless access providers, and the two major PTO operators.

The first layer is the relationship between the PTOs and the cellular/PCN operators. BT and Mercury are responsible for working with cellular providers to define interconnection tariffs for the carriage of those phone calls that are transported on their network. The PTO can set various interconnection agreements, although certain agreements have been tested and overturned after inquiries by Oftel. Much attention has been focused on the relationship between BT and Cellnet, of which BT is a majority shareholder, and between Mercury One-2-One and its parent company, Mercury. In both cases, regulators have investigated the need to restructure tarriffing arrangements to ensure cross-subsidization of wireless operations from wireline revenues does not occur.[7]

The second layer is the relationship between the four operators and the more than two dozen retail service providers. From the outset, Cellnet and Vodaphone could not sell their services directly to customers. This restriction caused the explosion of retail service centers in the late 1980s and early 1990s. Much like the traditional relationship between a wholesaler and a retailer, the retail service outlets would offer service packages directly to customers, often bundling the wholesale offerings of a wireless operator with the sale of a headset. Although the restriction has been lifted and no such restriction was placed on the two PCN operators, the market has been saturated by participants at the retail level; walking around London, it would appear that there is a retail outlet on each block of the downtown district.

The basis of the relationship between the service providers and the retail outlets is a mixture of incentive packages and discounts for service that allow the retailers to make money on increased sales. As of June 1994, just after the launch of the Orange network, the four wireless network operators were offering a total of 17 wholesale tariffs.

In the meantime, as this confusion established itself in the heart of the U.K.'s wireless access market, competition between wireline and wireless networks failed to emerge at all. In fact, while BT's wireline prices dropped significantly from 1985 to 1994, prices for peak calls on Cellnet remained unchanged during the same period (Escutia, 1996).

The retail layer has become a commercial bottleneck in the provision of services and, in a certain regard, a distraction to the potential emergence of price competition between wireless and wireline networks. Instead of providing the value added of differentiated rates and service information directly to customers, wireless operators have to offer a value-added to retailers in the hopes of motivating them to pass on the value-added and pricing components to customers. This takes us down to the final level of the service structure for wireless telephony in the U.K.: the relationship between the retail outlets and the customers. By the time the information reaches the customer level, there are literally thousands of permutations of tariffs, prices and packages. Most of them appear to be less differentiated by price than simply by marketing name and image.

From the perspective of regulation, this is a troublesome bottleneck. The institutions that have been identified as the critical component of further competition, namely the wireless service providers, are highly constrained in the breadth and depth of their strategic business choices. One of the results is high levels of customer churn, with customers changing providers at the rate of 20% to 25% per year. It becomes difficult, if not impossible, to institute strategies to reduce churn through the layer of retail service providers, themselves a source of churn as retail-outlets change favored wireless operators with alterations in tariff strategies.

This arrangement of retail competition and service oligopoly appears not to be sustainable from the business point of view, either. Reports are that few, if any, of the retail outlet chains that have grown up in the past decade are profitable at all. Although companies can run deficits for an extended period of time, creditors are not known for their infinite patience. Financial difficulties of certain retail chains make their purchase by the providers a likely trend.

But that is as it should be. Consolidation will allow service providers to define direct relationships with customers, and will help service providers in their efforts to offer a total solution to the customer. With wireless providers sandwiched uncomfortably between a dominant wireline carrier and a bewildering system of retail competition, the market bears only the illusion of competition.

From the technological point of view, the development of the market in the U.K. has been largely successful. Standards have been chosen and effectively rolled out to achieve national coverage. But the institutional arrangement for the provision of services will hamper future development of the sector. The U.K. has attempted three specific experiments to break through these constraints and break open the market for differentiated and competitive services. One of the experiments, CT-2, has largely been judged a failure; the other two, PCN and wireless local loop services, are both experiments in progress.

CT-2 (Telepoint): The Failed Experiment

CT-2, known as Telepoint in the U.K., was meant to work like a cordless telephone (thus, the designation "CT"). The difference was in the range; base stations would allow a 100-meter radius for wireless access, and various minicells would be constructed to offer pedestrians or local users basic telephony. In concept, the idea sounded promising. The execution, though, turned out the be no less than a disaster.

Four providers were licensed to offer Telepoint services:

- BYPS (Rabbit)—Barclays Bank, Phillips, and Shell (later purchased by Hutchinson Whampoa of Hong Kong and Hutchinson Mircotel subsidiary formed);

- Callport—Motorola, Shaye Communications and Mercury;

- Phoneport—BT, STC Telecom, France Telecom and NYNEX; and,

- Zonephone—Terrantic, British Technology Group, with various others.

The initial problem was in the technology. As Ian Channing (1994) put it in Mobile Communications International, "Unlike cellular radio, where there is general agreement on technology, there have been long and often acrimonious discussions concerning the appropriate technology for cordlessness, and these discussions are still on-going." The lack of a standard for the service was caused by a direct conflict between the investment of local manufacturers and the work of the European Community to codify a different standard for cordless communications based on an incompatible transmission technique.

A number of U.K. equipment manufacturers developed the CT-1 standard in the mid-1980s as an application to support cordless telephones, and CT-2 was a domestic effort to improve the quality of cordless access for home and business users. CT-2 was based on frequency division multiple access (FDMA) technology. At the same time, the ETSI was working on the development of the Digital European Cordless Telecommunications (DECT) standard, which was based on time division (TDMA). The inability to determine, early on, the standard for the provision of Telepoint services reduced the commitment on the part of service providers and equipment producers to fully develop the system. As a result, the three carriers that eventually rolled out the service instituted their own proprietary headsets and base stations, allowing for no interconnection and interoperation even though all of them were using the same frequency bands (864 to 868 MHz).

But the difficulties were more than technological. Some would argue that the very nature of the service was ill-conceived and that Oftel prematurely released Telepoint because of the pressure to increase the level of competition in the telecommunications marketplace (ECMI, 1990). Considering some of the drawbacks of the service, it would be difficult to argue with them. Even though it was priced reasonably at 5.50 ECUs per month (versus ECU 19.25 for cellular), Telepoint could not receive calls, only send them. Before Hutchinson's Rabbit service collapsed in late 1993, a paging capability had been built into the system to notify subscribers that an incoming call was being made so the subscriber could return the call, but that was certainly too little, too late. In addition, service coverage was spotty and there were a number of technical difficulties in placing calls from the perimeter of the advertised cell coverage.

Clearly, the CT-2/Telepoint experiment in the U.K. was a resounding failure,[8] but, towards the end of the CT-2 experiment, Britain opened a new chapter and began a new experiment: Personal Communications Networks (PCN). This ongoing experiment appears to be largely successful but, as already noted, a number of institutional constraints have arisen that threaten to choke the life out of PCN's future. This section reviews the successes and difficulties of the two PCN operators in the U.K.: Mercury's One-2-One and Hutchinson's Orange.

PCN: The Continuing Experiment

It is a fitting tribute to Marconi's legacy that the concept for and first implementation of Personal Communications Networks (PCN) occurred in his adopted homeland of England. The experiment presently being undertaken, at the cost of billions of dollars of investment and huge investments of time and risk, is perhaps one that

Marconi would have wanted to take himself: the establishment of a network of personal wireless communications to link an entire country together.

The experiment is no less than the construction of competing networks for telecommunications services. The success of this experiment will depend largely on how regulatory and economic institutions choose to support these newcomers to the sector, and how attractive PCN networks can be to consumers. If they turn out to be as successful as CT-2, then there is little future for them. Fortunately for the experimenters, this kind of failure is not at all likely considering the amazing growth in PCN since the launch of the first PCN network in 1993.

This is in great part because PCN is built on a foundation that avoids the mistakes of CT-2. The technologies of PCN are well established, and both networks presently in operation are implementing the DCS-1800 standards previously discussed. The services offered have already been determined as attractive, and follow the cellular model. From the perspective of international competitiveness, Britain has much more to gain through investment in PCN than it did in CT-2. "The Commission sees personal communications as an area in which Europe could lead the world," Ian Holt of Coopers & Lybrand wrote in a 1994 edition of *Mobile Communications International*. In turn, the U.K. is positioning itself to lead Europe and, in leading Europe, perhaps even lead the world.

But what exactly is the nature of the "PCN" model that is arising in the U.K.? Is it similar to the cellular model, as has already been suggested during the overview of the models for wireless access presented in the third chapter of this book? On the surface, there appears to be some substantial differences between the nature of the licenses and business operations of the PCN providers.

Licenses for PCN operators were different from cellular in three critical ways. First, PCN operators are allowed to provide their own millimetric radio links between base stations, sites and switching centers. Incumbent cellular operators were prevented from doing this until two years after the last of the PCN networks entered service. Second, PCN operators are allowed to share infrastructure, which could be especially useful in low-traffic areas or rural areas. Finally, PCN operators were permitted to establish their own retail-sales organizations to directly solicit business from subscribers.

Even with these major variations, the substantive differences between PCN and cellular in the U.K. have had more to do with the institutional constraints established by the existing market for cellular services. Alterations in licenses such as the ones mentioned have only changed the marketplace itself in a limited way, and much more will need to be done to dramatically alter the model for providing wireless access services in the U.K.

Mercury One-2-One

Initially, three PCN carriers were licensed. Two of the carriers, U.S. West and Mercury, decided to merge their PCN networks to form Mercury One-2-One, and the service was launched in September, 1993. Allen Hadden, head of Business

Policy at Mercury One-2-One, described the network at the time as "the first mobile telephone service designed, built and geared to the mass market."

When the service was launched, the coverage for the system extended to the M25 orbital motorway around London, a region which addressed a residential and working population of over 10 million. The catch was free local calling during off-peak periods. As Hadden (1994) described it:

> Mercury One-2-One offers the lowest priced mobile telephone service available in the U.K., and is the only service which provides free off-peak local calls to the fixed network for customers on the personal call tariff. This is equivalent to 234 days per year of free off-peak local calling, and is ideal for people wanting to make greater use of the service in the evenings and at weekends, and as an economic alternative to dialing using the fixed network.

The concentration on London area residents, those inside the M25 beltway, induced a reaction from Cellnet and Vodaphone, both of which dropped their premium cost for business calls in that region. Attempts to shore up customer bases among the two cellular providers proved to be somewhat unsuccessful at first, with many migrating to One-2-One or moving to the competitor during this period of pricing transition (Valletti and Cave, 1996).

But the pricing strategy was also meant to establish a competitive position for the company not only relative to the two other wireless access providers, but also against BT's wireline network. According to Mercury's estimates at the time, the share for radio base access in telephone services would leap from 2% to 15% by the year 2000 (*Mobile News*, June 1994).

The results of this innovative pricing strategy were, to say the least, unexpected. Parents began to buy the phones for their children and restrict their calling to the off-peak hours so as to reduce their use of phones in the evenings and on weekends. By establishing a small circle of friends, parents could give their children a really cool gift while simultaneously circumventing all call charges for their children. After about a year and a half, One-2-One, under increasing financial pressure, withdrew this offer in favor of more traditional cellular pricing plans. The pricing strategy was certainly an innovation, but the limited nature of the experiment and the difficulties it appears to have caused make it difficult to judge whether or not One-2-One will be able to sustain itself so that direct competition with wireline service providers becomes feasible. Some would claim that the company's heavy debt burden, and the imminent launch of other kinds of wireless access solutions, makes it a prime candidate for takeover or collapse in the years to come.

From an operational point of view, the regional strategy was the best starting point for a service that had yet to be tested in a large market. From the point of view of increasing the level of competition within the sector, though, the strategy meant one thing: The existing operators could relax. One-2-One was not going to be the threat they originally anticipated. By early March, 1994, Mercury One-2-One announced they had initiated a program of accelerated city center coverage in a number of major cities, including Liverpool, Manchester, Leeds, Sheffield, Nottingham, Bristol, Southampton, and Birmingham. By the middle of 1996, One-2-One's coverage still was less than half of the total U.K. population (Valletti and Cave, 1996).

Orange

In mid-1994, Hutchinson Microtel launched Orange,[9] the second PCN network to offer services in the U.K. The main differences between Orange and Mercury One-2-One are less in the nature of the service than they are in the initial strategy for infrastructure development—as well as in the initial performance, both financial and market penetration.

The initial coverage map for Orange looked dramatically different than for One-2-One. When Orange became operational, more than 40% of the country could access the network. To achieve this kind of coverage, Orange spent over 700 million pounds sterling by the end of 1994, and achieved its goal of 90% coverage of the country's population covered by the end of 1996. This national strategy will make it more difficult for the two cellular providers to "hide" by differentiating prices in specific regions of less competition.

PCN licenses permit the sharing of facilities with other PCN providers in order to promote increased and enhanced service coverage, but Orange has declined to work with One-2-One. In addition, Orange management consistently rejects assertions by some that a merger between Orange and One-2-One is in the cards as competition puts downward pressure on margins.

In fact, Orange has been working hard to highlight, through its financial and public communications, the differences between its network and One-2-One. When shares of the new PCN company were floated in March, 1996, Orange was quick to point out that the initial market valuation for the company was significantly higher than that of its only PCN competitor—some London city analysts claiming Orange's value to be almost twice as high, 2.4 billion pounds for Orange versus 1.2 billion for One-2-One (*Telecom Markets*, March 1996). Orange has also caught up to One-2-One during the course of 1996, signing more has more than 500,000 subscribers onto its network in a bit more than a year of existence.

Orange has positioned itself in the center of the market, slightly more expensive than One-2-One, but less expensive than the digital service offerings of Cellnet or Vodaphone. About half of those subscribers, though, have chosen Orange's Talk 15 plan, the lowest price option available in terms of monthly subscription cost offered by the company; 40% have chosen Talk 60, the second lowest price option (Valletti and Cave, 1996). One of the more potentially interesting innovations in the Orange network comes from the billing system, which offers subscribers costing by seconds, rather than by minute or half-minute increments.

It is clear that the company has made the investment necessary to compete within the cellular and wireless access market—that much is clear in the growth of the subscriber base and the impression in the press and among professionals in the industry that the company is on a strong, sustainable course. But Orange will have difficulty competing directly with the two major wireline service providers for the same reason that the cellular companies and One-2-One have not yet been successful in positioning their companies against BT: PCN is trapped in the cellular model of service provision, leaving it in a position where direct competition against wireline networks is a secondary portion of the business strategy.

U.K. PCN: Trapped by the Cellular Model

These two PCN providers have had to work within the constraints of the three-tiered model for wireless services. As wholesalers, they have established service offerings for retailers. Considering the overwhelming presence of retailers on the streets of London, the alternate strategy of bypassing the retail structure would be expensive and time consuming. Even though the option is open to them, the bulk of their work to expand their subscriber base will be through the retail outlets.

That places PCN operators at a disadvantage. The incumbent carriers have already been able to establish relationships; cellular companies responded in advance of the entry of these two PCN competitors by introducing new wholesale tariffs to capture the nonbusiness market (Valletti and Cave, 1996). They continue to respond as their digital services become more widely available and additional intelligent network features are built into the service offerings.

But this does not condemn PCN networks to the trash heap of history; it only makes their entry more difficult. With regard to their ability to compete with existing cellular service providers, it appears that PCN providers are doing well. "PCN is outselling Cellnet and Vodaphone by 5 subscriptions for every 2," *Mobile News* reported in January, 1995. "This fiercely competitive market may be producing losers as fast as winners," the article continues. "Although revenue from mobile operators are rising, revenue per subscriber is actually falling, which is putting pressure on margins. And the trend is set to continue for at least 10 years."

This is the real problem. The margins are disappearing just at a time when new entrants need margins to remain stable so that they can establish a presence in the market. It is a race between decreasing margins and the undercutting of the retail bottleneck. The chief concern is that a wireless provider or two will be lost in the process, thus returning the U.K. market to the duopoly provision of wireless services.

In the mean time, customers are more and more confused. "The choice to users is now so wide as to be bewildering, particularly to a new and perhaps dubious customer" (*Mobile News*, June, 1994). Many a new technology and technologically based service has been tripped up by consumer frustration and skepticism at its introduction, and there appears to be no mechanism in place to diminish the negative effects of the illusion of retail competition.

Wireless Local Loop: An Experiment (Re)Launched

After the collapse of the CT-2 Telepoint service providers, the remaining opportunities for fixed wireless local loop service seemed limited—licenses were made available above 3 MHz, but the inability of Telepoint providers to penetrate the local service market would have made any investor more than a bit skeptical about the potential of the technology in the U.K. market.

But one company, so far, has decided to go once more unto to the breach—Ionica, which launched its services in the middle of 1996 to high expectations, even after numerous delays during the course of 1994 and 1995. Ionica's technology operates at 3.4 GHz and uses time-division techniques; the company license permits it to operate throughout the country at that frequency, providing fixed telephony and interconnection to other telecommunications networks. At the end of 1996, Ionica's service was only available in a portion of the country, the East Anglia region to the south and east of London. The goal is to provide service to 75% of the country by the year 2000.

As is traditionally the case for new service providers, Ionica is attempting to compete on price, offering subscriber rates at a 10% to 15% discount from BT's prices (*Telecom Markets*, May, 1996). To make comparisons easier, the company also mirrored BT's tarriffing and plan structure for services for initial marketing and sales efforts.

The goal of the business plan that has circulated among the investment community is to achieve a 5% penetration rate, paying for network construction as customers are acquired in select regions. With customer acquisition costs associated with network construction as low as 10 pounds per person, according to chief executive Nigel Playford, that would appear to be an achievable goal, as long as the company can continue to compete on price. The selective nature of the coverage to date invites aggressive price competition from incumbent providers, especially BT, in order to avoid erosion of the subscriber base.

A variety of new wireless local loop licenses were also granted during 1996, including three in the 2 GHz range and another three in the 10 GHz range. The first set of licenses in the lower portion of the spectrum is meant to serve rural areas, with those in the higher range allotted for the provision of ISDN type services. The winners of the three ISDN licenses, Ionica, Mercury and National Telecommunications Limited (NTL), had not announced service launch dates as of the end of 1996.

Like the U.S., then, a swarm of licenses has followed on from the PCN and Cellular providers, few of which have actually had track records long enough to judge the real effectiveness of the business plans and impact on the competitive environment. Structurally, though, the existence of the licenses points to potential opportunities for new kinds of service provision—and because some of the institutional restrictions placed on the cellular providers do not exist, perhaps speeding the needed rationalization of the market into full service providers that connect with customers at both the retail and wholesale level.

Summary: The Market for Wireless Access in Britain

The institutional consensus shaped since the passage of the Telecommunications Act in 1984 has worked well enough. By constraining competition and ensuring that the U.K. remains on the forefront of European telecommunications development, there has been enough of an expanding market in wireless and wireline services to ensure that market participants have incentive to invest in new technologies and services. But the institutions that provided much of the energy in developing networks and services need to be brought into a new institutional relationship based on more direct competition for customers.

In order to achieve that goal, elements of the existing institutional relationship need to be pushed aside. In particular, the practices must be identified to help support wireless and cellular providers as they make the direct connection to customers and offer the value-added services that will help them grow. Wireline players, be they PTO or other licensed service providers (such as cable telephony providers), also need to be given a direct stake in the market for wireless services. As it presently stands, the development of Britain's telecommunications sector could be rocked by the failure of wireless systems unable to define and protect a market share. It is also threatened by what might be a lack of interest among providers to open their markets further.

In other words, the lines that have been drawn between PTOs, wireless access providers, cable telephone and television providers, and enhanced service and value-added providers, need to be erased. This convergence of institutions would fulfil the vision of Oftel as revealed in the speech quoted earlier: direct competition for basic services through the expansion of wireless networks and the convergence of basic with enhanced services to provide a fully competitive market with a number of differentiated service providers.

Some of the steps outlined in the third chapter of this book would have a positive impact on erasing those lines. The final section of our discussion on Britain's wireless access market focuses back on the proscriptions of strategic liberalization, and offers some direction as to how the market for telecommunications services can be broadened to ensure a competitive future.

STRATEGIC LIBERALIZATION IN THE U.K.

In the discussions of the U.S. case, the predominance of the cellular model for service provision was evident—it is no less so in the U.K., with new service providers adopting cellular-like strategies for pricing, network construction and marketing to their potential subscriber base. In a country where wireline deregulation and privatization has gone about as far as it can go, given the historical precedents of the monopoly PTO organization inherited by the Thatcher government in the early 1980s, it is noteworthy that competition on costs and kinds of services is not much farther advanced than in the U.S.—and, like the U.S., real evidence of wireline versus wireless competition is almost non-existent.

The cultural and technological protection approach to liberalization policy faces the same difficulties in the U.K. that it would in the U.S. The goal of transforming institutions so that they reflect competition principles requires less of a focus on what to protect and more concentration on how to induce change among corporate institutions by establishing conditions consistent with the public policy objectives of increased service penetration and sustainability. The opposite side of the political spectrum faces, in fact, a similar problem—relying on the technology alone ignores the true facts of the U.K. case and the political economy that grounds the path of U.K. telecommunications development. Institutional constraints will require a more deft approach, or else BT's competitive positioning as the company with the only complete service infrastructure in the country will win the day by default.

The competitive context is certainly different from the U.S., with the patterns of wireline duopoly and new cable telephony providers offering a different sort of forest than that of divestiture, the 1996 Telecommunications Act, and the torrent of interconnection agreements that have followed. But, in one regard, the two countries are the same: competition based on infrastructure, as well as service provision, does not exist, and cannot exist because of the constraints placed on the institutional evolution of service providers. Specifically, the level and nature of government regulation is such that the development of competition through wireless communications remains more a idea than a reality.

If Duch's thesis is correct and it is the pluralism of the U.K. political and social system that helps sustain the continuing restructuring of the telecommunications market, then a portion of that pluralism needs to be injected into the wireless access market as well—a market that has been oligopolistic, even with the introduction of the two PCN licensees. But what is the source of pluralism and a new institutional consensus for the U.K. case, and how would those sources differ from the energies in the U.S. that may transcend the local and national divisions inherent in the telecommunications policy framework?

The discussion of strategic liberalization in the U.K. focuses on the same three levels as in the U.S. case: the level of government intervention, the substance of government policy, and the direction of corporate strategy. In each of these areas, it is evident that the U.K. has taken some substantial steps further down the ideal road that has been mapped out in earlier chapters than has the U.S., but there is a significant opportunity for both sides to learn from their neighbors across the Atlantic pond and use the fertilizer of ideas to sustain the growing telecommunications sector.

The Level of Government Intervention

In the U.S. case, the discussion concentrated on how wireless access can help overcome the difficulties inherent in the split between national and local jurisdictions, simply because historical precedent has placed the locus of almost all regulatory matters in the federal government. By concentrating on diminishing the level of government intervention in the market for the provision of wireless access services, the U.S. has the opportunity to move competition forward much more quickly than would be the case for wireline services.

The U.K., as a unitary system, does not face as significant a challenge in that regard. It might even be said that, because of the concentration of economic and political power in London, that it faces the opposite difficulty—how to take the national focus on competition and make it a reality in various locations throughout the country. The level of government intervention, in that regard, should be targeted to the need to spread competition in service provision from the London business community out into the residential areas and smaller cities of the U.K., providing more than the one dominant (BT) and emerging (cable telephony) service provision options available.

From the perspective of wireless communications, the same two issues addressed in the U.S. example apply to the U.K.—license definitions and seamless service provision are both critical areas of concentration if the market for wireless access services are to become a force for the continued liberalization of the U.K. telecommunications marketplace.

Reduce License Definitions

Many of the restrictions applied to U.S. wireless access licensees are also applied to those in the U.K.. Licenses define the types of services provided, the technology in the case of DSC-1800 for PCN licensees, and the geographical scope of service provision. Like the U.S., the U.K. regulators have decided to apportion the spectrum as if different portions were appropriate for different services, although a greater degree of flexibility has been offered to wireless local loop licensee Ionica to bring other, higher bandwidth services to the market through its 3.4 GHz license.

The common law tradition in the U.K., though, and the evolving nature of license definition offer a different kind of opportunity for regulators. Because licenses can be altered through the consent of the regulator and the carrier, it is possible to rewrite licenses along the lines suggested, allowing cellular and PCN companies the flexibility to offer what they choose as the market evolves. The alterations in licenses will also likely increase the value of each license should one of the companies collapse under the weight of competition; an investor or group of investors could pick up the license and move in a different direction than the cellular model with a lesser degree of concern about the regulatory implications of the strategic decision.

This is also true with regard to the provision of retail wireless access services by the four main providers. All four have begun to move aggressively to market services directly to customers, from the first days for PCN providers, because the restrictions did not exist, and since the lifting of the restriction for cellular companies. This shift in licensing is an example of successful regulatory adaptation, and continued movement in this direction of consolidation is likely to be to the long-term benefit of the country's telecommunications marketplace.

Oftel has already stepped into a realm where the U.S. regulators have been loathe to go by allowing various companies to accumulate large portions of the spectrum in various bands. Mercury has an ISDN license, but also has partial ownership in One-2-One. BT has a fixed license to serve rural areas that could, in theory, be used to supplement local service provision if regulatory issues were resolved. The FCC has expressed more reservations about such license accumulation, in part incenting many companies to enter into the consortia bidding and operations structures apparent in the PCS auction in 1995 so that investments in wireless could be separated from other wireless and highly regulated wireline holdings.

The one possible complication in the implementation of this kind of policy would occur in the broader, European context. As wireless access services continue their growth on the continent, U.K. providers may be incented to develop stronger cross-country presences—moving beyond roaming arrangements to establishing new companies and consortia that offer regional, if not pan-European network coverage. License alignments with common frequency assignments for common types of services may help to facilitate that consolidation by simplifying network management. But that challenge has not dampened the progress of cellular providers in the U.S. that are attempting to create national services on the basis of collecting fragments and pieces in various frequency bands; this should not be viewed as a significant obstacle to European providers as 1998 approaches and monopoly restrictions are, in theory, lifted.

Nevertheless, the constraints that exist in this area serve no useful purpose and can not further the protection of consumers—there is no threat inherent in a cellular provider switching to fixed services or providing both in a competitive marketplace. Such restrictions can only serve to hamper the evolution of services in local communities, where penetration of a variety of wireless service options would be for the betterment of competition.

Seamless Service Provision

The rationale for separating wireless and wireline service provision in the U.K. is very similar to that expressed in the U.S. It is the problem of cross- subsidization, of unfair boundaries to interconnection on the part of BT or other incumbent wireline providers that exercise a bottleneck control over local access. These are very real problems but, in a market where full facilities based competition should provide a broad range of access options, bottlenecks are likely to disappear, and the incentive to use remaining bottlenecks diminishes as the value of completely interconnected networks increases. In addition, the possibility of cross-subsidization should, in theory, also decrease as competition between wireline and wireless networks increases, because profits from wireline and wireless services will vary from location to location, with one outstripping another for a company in various jurisdictions. It might even get to the point where wireless services are used to "subsidize" the wireline services traditionally regulated by the government because of its potential as a source of monopoly profits.

A regulatory focus on interconnection will continue to be critical in the U.K. context, just as it will be in the U.S. Nevertheless, the protection of interconnection as a value for the national community does not necessarily require that restrictions on bundled services and seamless internal operations of company networks be maintained. Part of the challenge will be to define where that balance point lies as each company plots its investment strategy and sees its fortunes rise and fall with the push and pull of the market.

The Substance of Government Policy

In the U.S. case, the substance of government policy required to implement a greater liberalization of the market for wireless access services concerned two issues: co-location of facilities and the role of auctions and incentives. In many ways, these two points are particular to the U.S. context, especially the strong local jurisdiction over facilities and the strong incentive to identify new sources of government revenues. In the U.K., site locations and infrastructure facilities have been a difficulty, for sure, especially because local communities are facing two buildouts at once: towers and cell sites for wireless access as well as cable telephony/television systems. But the unitary nature of the U.K. political system offers some clearer direction on resolution of these issues, such that the challenges do not appear to be as great for service providers on the European side of the Atlantic.

The second issue, the role for auctions and incentives, is one which should be considered in light of the U.K.'s issuance of further licenses for various wireless access services. The critical point for discussion, though, may have less to do with the U.K. and more to do with the U.K.'s position in Europe and in the world as a driver of telecommunications service development and innovator in the area of new products and services.

A Role for Auctions and Incentives

For the evolution of the wireless service market in the U.K., the bottleneck right now is at the retail level. Institutions have been constructed and put into place which restrict the development of a direct relationship between the company and the customer. Retailers do not need to differentiate themselves by operator affiliation at present because there are a limited number of providers, a limited number of tariffs being offered, and they are the only ones with direct access to the broad customer base.

Breaking up that bottleneck requires more operators and fewer retailers. If the retailers are not profitable, then the "fewer retailer" problem will correct itself over time. But, as is always the case with wireless services, the U.K. government must step forward and begin the process for a more extensive licensing scheme so that there are opportunities for additional providers to enter the marketplace.

The U.K. has not used auctions for spectrum licenses, choosing instead to use a traditional "beauty contest" method for evaluating the capabilities of various investors and service providers. The success of the system to date contrasts with the U.S. lotteries of the early 1980s, which only succeeded in infuriating those who wished to participate (though certainly pleasing the winners). The success of the auctions in the U.S. was not in the revenue gained for the government, although that is a significant benefit; it was the valuation of the spectrum. By valuing the spectrum, it becomes possible to make sound public policy and business decisions based not on the whim of regulators but on the needs and interests of the market. By using the beauty contest method, the U.K. lacks an important valuation scheme for spectrum.

The introduction of a valuation scheme provides a range of planning benefits, not the least of which is to provide direction for businesses that would consider entering the market for wireless services. The opportunity to open other frequency bands through an auction would create the preconditions for increasing the number of providers.

It would also likely serve to decrease the number of retailers. As margins are squeezed and the value added of customer service becomes more and more a critical element of profitability, operators were vertically integrated from the retail to wholesale levels, and existing franchises will become ripe for purchase by new or existing licensees. Auctions will likely increase the number of providers and decrease the number of retailers, and also provide a foundation for the implementation of other potentially beneficial policies, such as an pioneer's preference program to link the U.K. markets with European and global technological innovation.

The U.K. does have a vibrant local market for electronic components and equipment for wireless systems. But there is no single provider that has the international reach of a Motorola, Lucent, NTT, Ericsson, Siemens, or Nortel. The U.K. has been able to turn this disadvantage into an advantage by establishing an advanced market for services. Creating a framework for increasing the number of providers in the marketplace will do much to sustain that competitive lead, but there is another piece of the equation that U.K. regulators should consider: putting in place a technological preference scheme to ensure that Britain retains its position as the most innovative market in the world.

The pioneer's preference that the FCC tied to the PCS auctions in early 1995 were fraught with political difficulties. Between court challenges, extensive comments by industry participants, and FCC decisions, the level of frustration was high, obscuring the benefits of what is a very good idea. By establishing a preference scheme for the development and implementation of new technologies, government can help to speed the benefits of new innovation to the marketplace.

For the United States and its wide variety of equipment manufacturers, the opportunity to work with a number of large and smaller companies in defining what is truly innovative has its advantages, but it also has its price. There is no political will to look outside of the country to solutions which may be more appropriate or just as worthy of recognition and investment. As a smaller country and integral part of the European system, the U.K. could establish itself as a judge of emerging technologies and the best trial ground for these new technologies.

As licenses are prepared for auction or for distribution through a beauty contest mechanism, Oftel should take proposals on solutions for the next generation of wireless technology, most likely two-way limited interactive multimedia systems that can directly compete with wireline networks. With an auction system in place, or an appropriate valuation of spectrum based on an independent assessment of the potential future returns from the license, companies interested in playing a role in the process will know the value of the reward and will be able to allocate research and promotional resources accordingly. Additional licenses can be provided to those systems considered most innovative, either at a discount to the auctioned price or free for the use of the licensee.

The licensee would then be asked to produce a certain percentage of the components for the system in partnership with local businesses. In addition, the depth of experience in providing wireless services throughout the country would offer a vibrant talent pool of human resources for the new entrant to use in establishing broad-based services, should they choose to do so.

At worst, such a system rewards technological innovation and brings an awareness of and opportunity to invest in the innovation to the participants in the British market. At best, such an incentive structure ensures an ongoing investment in the U.K. telecommunications marketplace, both in services and in equipment manufacturing.

The Direction of Corporate Strategy

The direction of corporate strategy during this period of global transition in the telecommunications marketplace is likely to depend highly on local context and national frameworks for infrastructure development. In the case of the U.S., the openings in the area of fixed wireless service and the long-term squeeze in opportunities for reselling and outsourcing were the critical areas of discussion. Reselling, a significant portion of the U.S. market for wireless access, probably faces a similar future of decline in the U.K. as telecommunications providers are forced to better align capacity with market demand. Outsourcing of billing and other kinds of administrative support, as discussed in the U.S. case, is likely to be significant, but of a different nature than in some developing countries, where this kind of support could be thought of as a very significant barrier to entry.

Fixed Wireless Services

The role of fixed wireless services, though, deserves attention as a point of comparison and opportunity for corporate strategy in the U.K. The U.K., unlike the U.S., has already launched fixed wireless services—they have been launched twice, in fact, once unsuccessfully, and once too recently to judge. The U.K. is not, with certain service patterns already developing for fixed services and their potential overlap with wireline offerings. For the U.K., the implications of fixed wireless services will likely be elaborated differently on two levels: basic telephony and value added, broadband service provision.

For basic telephony, the most likely competition for local fixed-link services will occur between the cable telephony providers and BT. But the cable providers are not using a single platform to provide that competition, and have been extending a phone line separately to the home and treating the telephone system as independent from the cable system. Now there are three wires to the home (cable telephone, cable television, and BT), which certainly is an immense investment in plant that would be difficult to replicate for other entrants. At best, such a system can support two competitors, but, as the Duopoly Review made clear, there need to be more than two competitors for there to be a competitive marketplace.

The pattern represented by Ionica's corporate strategy seems to dictate that fixed services will largely be used as an entry strategy by new providers, rather than as a play to extend service franchises by existing providers. Cable providers, for instance, have yet to receive the opportunity to use fixed services to move outside their franchise area and broaden telephony offerings, even though they have applied for licenses that would permit them to do so. The alternative, purchasing fixed systems as they are developed or merging with them to compete against BT, is likely to be the only option for many of the newest market entrants as they struggle to compete with the ubiquitous marketing and network presence of the dominant provider.

In addition, even though the market for fixed wireless services collapsed after the failure of CT-2 (Telepoint) services, Telepoint deserves a second chance. The main reason has to do with cost. Wireless services in the U.K. are still predominantly focused on mobility. As in the United States, PCN providers will be challenged to serve the automobile mobile market as much as the pedestrian and residential marketplace. These pressures increase the capital cost of the network and the cost to consumers, thus delaying the time when wireless and wireline networks can directly compete with each other for market share. In that regard, the U.S. is much more a blank slate, with companies concentrating largely on cellular-type applications to meet the continuing demand for automobile-based telephony, rather than residential or pedestrian applications.

It also has to do with spectrum. As long as the licenses lie unused, opportunities for service provision remain unrealized. Commercial opportunities will emerge from these properties, even if it is only paging or some sort of wireless data service that forms the basis of additional service offerings.

Broadband services is an area where the U.K. market is wide open, in great part because cable providers have only begun to make its services broadly available. The emerging options for entertainment also include satellite services like BSkyB—although satellite providers have largely concentrated on broadcast, entertainment services are often an important part of an overall package which, when combined with services like high-speed Internet access, will offer important marketing and sales opportunities for telecommunications service providers. The market in this area, therefore, is already being squeezed, in part, by the evolution of the television markets in the U.K.

Wireless access will offer a significant opportunity for convergence, a single platform that can be used to bundle entertainment and certain kinds of two-way communications, either through basic telephony or through ISDN type services using the presently available licenses. Because it is open territory, the political hurdles to implementing wireless convergence are likely to be less significant as well. On this second level, then, companies like Mercury or individual cable providers will be well served with investments in this evolving portion of the marketplace.

CONCLUSIONS

Throughout this chapter, the issue of institutional consensus has been emphasized as a potential determinant factor for the success of an ongoing liberalization program. In many ways, the importance of the pluralistic structure of the political and economic system has been assumed as the critical consideration for policymaking in support of further liberalization in the U.K. It is only by connecting corporate strategy and institutional evolution to appropriate public policy that the foundations for the sustainable development of the telecommunications sector can be set.

Companies from the U.K. market have a greater depth of experience in working in an environment with multiple service providers than U.S. companies, and executives pay greater attention (publicly, if not in the details of corporate strategies) to the potential competition between wireless and wireline infrastructures than do their U.S. counterparts. Both of those factors make a difference in the potential implementation of competitive principles in the wireless access marketplace. What the discussion does make clear, though, is that the potential for wireless liberalization is significantly greater than wireline liberalization, if only because the duopoly has not given way to facilities- based competition on a national scale, only on a regional or local scale, and then largely on price rather than on the potential combination of carriage and value added services. Pushing the market further, and sustaining multiple providers on a national scale, will require a greater investment in wireless access as the foundation for government liberalization policy and corporate strategy.

One thing should be made clear. These proscriptions set a dynamic in motion that could be unpalatable for one specific reason: Telecommunications companies could fail. Saturation of the market will be achieved as penetration rates increase, and companies will collapse under the competitive pressure. In any other industry, this would not be considered a problem; it would be perceived as the natural consequence of market development. But in telecommunications policy, there is still little understanding of the consequences of such a position, in either the short or long term.

That would be the boldest experiment of all, one for which the U.K. is already largely prepared. The question for the next decade is clear enough: Will the U.K. embrace the experiment by implementing a policy of increasing liberalization in the wireless services market, or will the present institutional arrangement favoring wireline services prevent the market's evolution? Considering the strengths of the open political system that U.K. has established in the post-World War II period, it is likely that further changes will be made and that the gains of the past decade will certainly not be for naught.

ENDNOTES

[1]   Garrison described the decision to privatize BT as "the unanticipated outcome of conflict between different government economic policies" (Garrison, 1988). Breaking up BT in a fashion similar to the break-up of AT&T in the United States was not a possibility because of the opposition of both BT management and unions, whereas privatization was seen as a lesser evil. There was even a lack of consensus about the privatization of BT among Conservatives; Harold Macmillan, the former Conservative Prime Minister, likened the privatization of BT to "selling off the family silver" in a 1985 speech to the House of Lords. For more information on the politics behind the decision to privatize BT, please see John A. C. King's remarks, entitled "The Privatization of BT in the U.K.," to the World Bank Symposium on Telecommunications Policy, 1990.

[2]   Mercury was initially a joint venture between C&W, British Rail, and Barclays Bank. Mercury is presently a wholly-owned subsidiary of Cable & Wireless, and, as of early 1997, was in the process of merging with three major cable providers—NYNEX CableComms, Bell Cablemedia and Videotron.

[3]   BT was brought under strict price regulation, also similar to the utility pricing used by American regulators (commonly known as RPI-X). The pricing regulation allowed BT to raise its prices by a margin lower (or, less often, higher) than the increases in customer prices, with the goal of incenting the company to increase its efficiency in providing services.

[4]   Paschal Preston, "Competition in the Telecommunications Infrastructure; Implications for the Peripheral Regions and Small Countries of Europe." *Telecommunications Policy*, August 1995. On p. 267, Preston wrote:

> It was not an accident nor merely a question of ideology that the Thatcher government pushed for the opening up of EC telecommunications markets in the wake of BT's privatization. Its active promotion of a pro-competition and free-trade regime in telecommunications markets elsewhere was clearly based to a great extent on perceptions of BT's comparative advantages compared to other PTOs, the economic advantages and benefits that were perceived to flow to BT, the U.K. national economy (and government revenues) as a result. This is made quite explicit in a number of U.K. policy documents on the 1980s.

[5] Marconi founded the first wireless company in the world, called the "Wireless Telegraph and Signal Company," on July 20, 1897. Before that time, he had serviced various government institutions and attempted to establish some private wireless telegraph services. But the Telegraph Acts of 1868 and 1869 (which entrenched the government monopoly of telegraph transmission in Britain) prohibited the company from instituting a competitive inland message-carrying service, forcing the company to look to "ship to shore" communications. Because the U.K. had the largest naval presence in the world, the market for the British government was enough to sustain the new company.

[6] For more information, see W. J. Baker, *The History of the Marconi Company*, (London: Methuen, 1970) pp. 28 to 30. Baker pointed out that the conflict between the emerging service of wireless communications and the government wireline network was as much a personal as an institutional conflict. The Chief Engineer of the Government Postal Office (GPO) and Marconi were often at loggerheads during the first decades of this century.

[7] The most notable complaint about the cross-subsidization of services came from Talkland International (U.K.) Limited, an independent service provider company retailing airtime on cellular telephone networks. Talkland filed a complaint with Oftel in 1992, and Oftel's findings were announced in 1994. Oftel concluded that unfair cross subsidies were provided by BT to Cellnet and by Mercury to One-2-One. Oftel threatened an order to require immediate compliance from BT in removing the cross-subsidy. Because of Mercury's position in the marketplace, Oftel did not deem these cross subsidizes as inappropriate for the starting of operations, but decided that they should continue for no longer than one year from the announcement of the finding. For more information, see "Fair Competition in Mobile Service Provision," Statement by the Director General of Telecommunications. (London: Oftel, May 1994). These directions were revoked after both companies were found to be in compliance in October of the same year.

[8] See Adrian Morant, "The Growth Sector in U.K. Telecommunications," *Cellular & Mobile International*, May/June 1993.

[9] The origin of the name choice for the network is worth a footnote. Marketing Director Chris Moss, in a speech during the press event to launch the network, explained:

[Orange] could never simply be just another mobile phone. It had to be an attitude of mind. A new frontier-breaking, wire-free world, a distinctive new service. Most of us were uncomfortable with the Microtel name. It fitted well with our technical profile of small telephones. But we researched it and we were staggered to discover that over 1,200 companies had either "micro," "tel" or "phone" in their name. It was too reminiscent of microwave and Microsoft. It was a me-too image and this was no way to convey a wire-free world.

Creating an Open Communications Environment: Strategic Liberalization in Russia

The struggle for political power and authority in Russia is based on dynamics surprisingly similar to those in place during the height of Soviet control—a tentative balance has been struck between various sectors and interest groups, each defined by its own economic and geographic affiliations.[1] That balance has sustained Boris Yeltsin and his team of government ministers since the fall of the Soviet government at the beginning of the 1990s. Even after Yeltsin's generation of political leadership fades from the scene, the pressure points for Russian society will remain the same, centered in the country's unique sense of itself as an historical actor in an evolving global marketplace.

Part of the problem in reading Russian history, though, is in the determination of what represents "authority" and what reflects acts of "power." One of the commentators on Russian events most commonly associated with the discussion of authority in the Soviet and Russian context is Thane Gustafson, whose particular interest in the post-Soviet world has taken him further into the analysis of Russia's oil and natural resource industry—a sector that has its parallels to the telecommunications industry, if only in terms of the scale and high-profile nature of the investments made. Gustafson pointed out on a number of occasions (1989) that the words for *power* and *authority* are used very deliberately in the Russian language—the root for the word *authority* actually has religious connotations that would connect the concept with "higher authority," while *power* is often used in the sense of "brute force."

The tension between power and authority will continue to mark Russia's development. The locus of authority, for instance, has shifted a great deal since the fall of the Soviet Union, with regions stepping forward and taking control of their own economic and political destiny. As the *Economist* reported in March, 1995:

Political power is eddying out from Moscow. Local governments now control over 80% of public spending on welfare. They have done most of Russia's privatizing. They still have the main say on who owns the country's land. They decide how fast the next stage of reform will proceed on the ground. That means they can influence how much the poor are cushioned from the harsh side of reform and thus, to quite a degree, how they will vote in parliamentary elections at the end of the year.

Only a year and a half later, the reports were that, with Alexander Lebed in charge of the military and Victor Chernomyrdin installed for another term as Prime Minister, "authority" to get things done may be returning, in part, to the center. But that coalition quickly disentegrated after the election and a new sense of instability at the center has settled in. Next year's reporting will certainly reflect something different, but the terms of the debate will remain the same—who has authority to do what, and where is power, or brute force, likely to be used to further specific political and economic agendas?

The telecommunications sector, as a part of the evolving political and economic landscape in Russia, will depend on how the forces of authority and power align in the years to come, but the past of the telecommunications sector is certainly a reflection of how those patterns have evolved. Russia has embarked on a wildly chaotic telecommunications development path. And it is a real possibility that the sector will establish business and regulatory patterns that will negatively impact the overall development of the country.

The question becomes: What institutions, if any, have the authority required to set the terms of the debate and contribute to the development of a stronger telecommunications sector for the country? And how can investment in specific kinds of technologies, when coupled with appropriate levels of regulatory and economic reform, help to further the evolution of these corporate and public sector institutions?

The preceding discussions have made it clear that telecommunications development is inextricably linked to access—the ability of individual subscribers to gain access to the telecommunications network. Investments in telecommunications products and services that do not contribute to individual and business access do not, per se, appear to lead to the sustainably competitive development of the telecommunications sector. The discussion of the Russian case returns to that theme, primarilly because the events of the past 5 years indicate that this proposition is more than a theory. Investments in large-scale infrastructure projects have, in great part, syphoned off needed dollars that could have been invested in more than cream skimming operations in various localities throughout the country.

It is the localities that are most critical in the discussion of Russia's telecommunications development; in great part, that is where the only authority exists to drive telecommunications development today. For that reason, a grass roots approach which has, as its goal, the linking of various regions into broader networks for development probably would offer the greatest possible reward, both to investors and to individuals involved in the public policy process (Kazachkov, Knight and Regli, 1996).

On a national scale, therefore, Russia needs to establish an open and unified communications space for telecommunications development. This unified communications space would consist of astandard economic and technological pattern for the integration and interconnection of communications networks and asystem architecture appropriate to the needs of the country, specifically one founded on wireless communications technologies. Through the support of the unified communications space, government and private sector managers can ensure healthy grass roots investment, which contributes to the overall economic development of the country. By using the guidelines of the strategic liberalization policy to establish this unified communications space, Russia can do much to keep the telecommunications development on track for the future.

This chapter opens with a detailed examination of Russia's present telecommunications infrastructure, including a brief discussion of the Soviet period and its impact on the present infrastructure. Next, the market for wireless communications is examined and compared to developments in the U.S. and the U.K. detailed in previous chapters. In closing, the discussion returns to the sustainable competitive development of Russia's telecommunications infrastructure and how the lessons of strategic liberalization might guide the level of government involvement in the sector, as well as some specific elements of corporate strategy and government policy.

UNDERSTANDING THE
RUSSIAN TELECOMMUNICATIONS INDUSTRY

Any discussion of the future of Russia's telecommunications infrastructure must take into account the uniquely Russian social realities; to evoke the traditional image employed by writers like Turgenev and Dostoyevsky to describe the country, there will always be more than one horse driving the troika. Understanding the Russian telecommunications industry, much like understanding the country's history, requires a keen focus on how individuals and institutions drive, both in concert and in conflict, the development of the sector.

Sustainable development, especially in the Russian case, therefore cannot mean a single plan, or a central focus on how development is to proceed. The definitions and descriptions appropriate for policy and corporate strategy can not ignore the strength of the many different efforts that have emerged on the local level. How those local efforts got started, how a space opened up for legitimate authorities to pursue local telecommunications development throughout the country— that is the key to understanding both the nature of the industry and some of the policies and corporate strategies with the most likelihood of success.

The history of telecommunications development in Russia stretches back as far as the 1840s, during a period of great investment and infrastructure growth in the country. Ironically, the investment in railroads, telegraph lines, and modernizing infrastructure did little to hold the country together until these assets were assembled by the force and power of the Stalinist state. Even so, Russia has always been threatened by fragmentation, even during the enforced stability of the Communist period.

History would appear to indicate that, in some measure, bad public policy and infrastructure development programs were responsible for the lack of cohesion during the late 19th century. The links of communication constructed to bring the country together actually tore the country apart. When the first railroad was put into place in Russia during the reign of Nicholas I, for example, it is said that Nicholas himself resolved a dispute among his engineers concerning the route for the train by laying his sword on a map and connecting with a straight line the points between St. Petersburg and Moscow. In the words of one Western historian, "He drew a straight line between them which the railroad obediently followed, even though the line sometimes led through swamps and passed nowhere near the townships it served" (Crankshaw, 1976).

The political interests were the first to be served by telecommunications technologies as well. Telegraph lines were first built in Russia in the 1840s and, in 1843, a telegraph line was extended to connect the summer and winter residences of the Czar in St. Petersburg. Soon after Alexander Graham Bell patented the telephone in the United States, a number of Russian engineers worked to modify that invention with new kinds of microphones and switching apparatus. The work of Russian engineers lead to the construction of the first telephone exchanges in Russia in the early 1880s; by 1885, urban telephone exchanges were in place in St. Petersburg, Moscow, Odessa, and Kiev.

But it was during the Soviet period that the first real investments in telecommunications infrastructure were made. The following discussion shows, though, that ideology and political purposes subverted the healthy development of the sector, leaving marked patterns of power and authority that have been picked up as present day assets to drive infrastructure investments.

Telecommunications During the Soviet Period

Many of the present difficulties faced by the participants in the telecommunications sector are due to the market inefficiencies that generally characterize state ownership of a telecommunications provider. State-owned telecommunications companies tend to be underinvested because governments manage them as cash generating activities, rather than as businesses to be developed. Profits that could be reinvested for long-term returns are diverted to cover immediate government expenditures.

Another distortion that affects all state-owned telecommunications providers is overstaffing. Administrators are unable to cut staff when appropriate because of the political problems of government layoffs—in some cases, managers do not even know how many employees they have on the payroll and what each employee is being paid. Bloated staffs make payroll expenses higher than they might otherwise be, thereby making the return on investment correspondingly smaller and less attractive.

Both of these difficulties marked BT's development before privatization dramatically changed the market environment in the U.K.—but, even so, the difficulties that arose from the U.K. pattern of telecommunications development pale in comparison with those of Soviet Russia. The problems do not end with these two systemic difficulties; Soviet underinvestment in telecommunications was a result not only of general tendencies of state-owned telecommunications operators, but also of distinct factors related to the lack of market signals in the Soviet economy.

In the Soviet case, there were additional factors discouraging investment in the telecommunications sector. First was the traditional Soviet aversion to light industry and service industries. Throughout the Soviet era, the high-status industries were large-scale, heavy industries such as steel, industrial equipment, and military hardware. Industries such as telecommunications, with less tangible economic benefits, were starved and allowed to grow at a very slow pace.

A key impediment to telecommunications investment concerned the output indicator method of evaluating industrial performance. Heavy industry lends itself to tangible output measures in tons of steel, coal, and oil, thousands of tractors, and so forth—while only the hardware side of the communications industry is amenable to such metrics. The real value to the consumer of a telecommunications service, namely the connection itself and the ability to communicate, is difficult to measure in the absence of meaningful prices. The prices in which the Soviet authorities enumerated were uncorrelated with the true value of goods, so there was no way for authorities to learn how valuable communications services were to the national economy (Neuman et al., 1995).

Another distortion caused by the output indicator method was its irrational application. Indicators such as "kilometres of cable laid" or "apartments wired for telephone service" were declared, regardless of whether the cable was ever put into use. Similarly, the target for apartments wired for telephone service was calculated regardless of whether the apartment dwellers actually received telephone service. All that mattered was that the cabling had been installed.

There is also the harsh reality of Soviet totalitarianism to blame for the configuration of the telecommunications network. After the Bolshevik revolution of 1917, the purpose and structure of the telecommunications networks were transformed into a portion of the totalitarian system that was to be developed by the Communist Party. Along with a total control of the press, the telecommunications infrastructure became intimately tied to the political system.

The architecture for the country's telecommunications networks reflected the centralized power base in Moscow. As technologies advanced, the political tension between the demand for access to communications and the Soviet government's need for absolute control increasingly were at odds—a trend which likely contributed to the collapse of the Soviet system in the 1990's.

The center for the political control of the telecommunications network was the Ministry of Communications, usually abbreviated conversationally as Minsviaz. During the later Soviet era, Minsviaz was responsible for all of the telecommunications networks throughout the country, including the construction, maintenance, and operation of all television and radio broadcasting facilities, and the post offices. In that regard, the Minsviaz was a typical PTT administration, controlling all of the communications resources throughout the country. The communications architecture reflected the political structure: All the telecommunications lines led to Moscow. As Robert Campbell put it in 1988:

> The Moscow UAK (Russian for uzly avtomactichgeskoi kommuntatsii, or automatic switching exchange) is the only transit exchange connected to all the other 14 tertiary transit exchanges. Apparently the tertiary offices all use foreign equipment, mostly the ARM or MT-20 exchanges. In addition to serving as the international gateway for telephone communications, Moscow also serves as the gateway for the telegraph and telex international connections. (Campbell, 1988, p. 12)

Because of the hierarchical communications structure and the compartmentalization of the economy and society under the communist system, the networks lacked integration and connectivity. On top of that, the service was "bad in numerous dimensions," to use Campbell's understatement.

But even under central planning there were some proto-market forces at work to expand the capabilities of the Soviet communications networks. This fascinating development is detailed by Campbell (1988):

> Minsviaz received a new charter in 1968 that gave it responsibility for the whole system, and rights to check branch systems for economic justification, conformity to Minsviaz standards and compatibility for connection to the utility network. Branch systems were expected, where possible, to lease lines from Minsviaz, or where appropriate links did not exist, to finance them cooperatively with Minsviaz and other ministries, with ownership going to Minsviaz. This process has always involved a great deal of conflict and coordination is still rather incomplete. The Minsviaz case stresses the advantages of universal access, standardization, compatibility and cost saving. Minsviaz argues that the trunks in branch system are ineffectively used—one source says that they are half as heavily used as Minsviaz lines—and do not meet quality standards. The client ministries believe that their needs are different, that Minsviaz does not serve them effectively, that they can do the job better than Minsviaz. The jurisdictional battles are fought out in an interdepartmental coordinating council, and it is not surprising that some of the bureaucratically powerful ministries manage to get their plans approved. Minenergio, Mingaz, Minneft (oil) and the railroads, all of which can argue for special circumstances, maintain substantial independent systems. For example, the oil and gas ministries have extensive operations in areas where the Minsviaz network was not devel-

oped. Minenergio has a far flung network of facilities that it has to keep coordinated in real time. The railroads have a long tradition of operating their
own telegraph and telephone system, and a distinctive combination of signaling, telegraphic and telephonic communications to handle. In all these
cases, the non-Minsviaz agencies sometimes make their own equipment,
order it from a domestic supplier in competition with Minsviaz, or import
it. For the BAM, the railroad ministry was able to acquire a special communications system, built by a foreign firm. (Campbell, 1988, p. 75)

It is important to put these events in historical context. Leonid Brezhnev had
consolidated his power, and Communism entered into a phase to which historians
have attached terms like "stagnation" and "gerintocracy." With the higher levels
of power interested more in consolidation and stability in reaction to the chaos of
Stalin and the inconsistency of Nikita Krushchev, it is apparent that much was going on under the surface. These competitive battles sound almost identical to the
bypass arrangements made by larger firms in countries like the U.S. and U.K. during the 1960s and 1970s, indicating that the facade of Soviet homogeneity belied
the real struggles for resources and influence being carried out behind the scenes.[2]

The fragmentation of networks, even during the Soviet period, was matched
by a fragmentation in the Research and Development apparatus of the Soviet system. Minsviaz itself had a very weak R&D base, with most of the resources being
focused in the Scientific Research Institutes, which were connected to the USSR
Academy of Sciences, and what are known as the VPK (military industrial complex) ministries, such as defense and security. In addition, much of the actual production occurred outside Russia in a number of Eastern European countries, most
notably in Eastern Germany, Yugoslavia, and Czechoslovakia (Neuman et al.,
1995).

There were some attempts to bring the pieces together into a unified network
but, as has been the case in the developed world as well, those efforts met with
little success. In the 1970s, a program called the Unified Automated System of
Telecommunications was initiated to improve the overall performance of the Soviet telecommunications network and to bring together into an efficient whole the
diverse networks that had been constructed. The results of the UACS program
were limited—even up to the present day, the former departmental networks still
exist, and they remain poorly connected to the rest of the Russian communications
network (Campbell, 1988).

The Rise of the Russian Republic
and the Russian Telecommunications System

This fragmentation was waiting to burst into the light of day as the era of glasnost
waned and the rise of the constituent republics began in the 1990s.

The initial thrust of Soviet telecommunications reform involved trade and
technology transfer. The Soviet Union sought the opportunity to gain access to

higher levels of technology exchange than formerly through the General Agreement on Trade and Tariffs (GATT) and bilateral agreements with various countries. Minsviaz played an important role as a contact point for foreign firms, but it was the VPK ministries and the individual research institutions that drove much of the production.[3] In the early 1990s, there were a number of smaller equipment sales—for example, Ericsson registered its first sale of MD110 digital PBX systems in July, 1990—but nothing so substantial as to indicate that the Soviet central government was contemplating the complete reconstruction of the public telephone network.

The first wave of telecommunications transport agreements focused almost exclusively on international carriage. In 1988, IDB launched a service that used privately leased satellite circuits for video and international services. Sprint International concluded an agreement with the Soviet Ministry of Communications to operate an electronic mail and videoconferencing service, using INTELSAT transponders and mostly focused on international usage.

While the discussions and national negotiations were taking place, a number of companies began to make contacts with local administrators and commercial concerns, especially in the larger urban centers of Moscow and St. Petersburg. These negotiations led to some of the first substantial foreign investments in cellular telecommunications networks throughout Russia. In November, 1990, U.S. West signed an agreement to take a 40% stake in a joint venture to construct Leningrad's cellular telephone network. The venture included the Leningrad City Telephone Network Production Association (LCNA) and the Leningrad Station of Technical Radio Control (LSTC), and operates today as Delta Cellular Communications. For Moscow, the city government and the MGTS (Moscow City Telephone Company) founded Moscow Cellular Communications with U.S. West and Millicom.

The provision of the cellular licenses in Leningrad and Moscow in 1991 are a watershed event in the jostling between local and national interests in the telecommunications sector. As one commentator characterized the situation in the early stages of the development of the Moscow system:

> When establishing a Moscow Cellular joint venture in Moscow, U.S. West was told by the Ministry of Communications that there would be two standards for cellular in Russia: NMT-450 and GSM. You can imagine U.S. West was a little surprised when Plexys, a U.S. manufacturer of cell-site equipment, was able to make a deal with a Russian defence contractor and get access to some frequencies that had been controlled by the military in the 800 MHz band for an AMPS system.

> When the contract for Moscow Cellular was awarded, the company thought it had frequency approval, which was a reasonable assumption under normal circumstances. It turns out that frequency approval is a bit more complicated. Historically, it has been controlled by the military and the KGB. There was no mechanism for civil approval of frequencies. (Leff, *Cellular and Mobile International,* March/April, 1993)

These administrative and legal gaps created opportunities for different communities of interests to assert themselves, mostly following along the fault lines that were an evident part of the Soviet telecommunications system. Basically, the cities appropriated the bandwidth directly under the nose of the Ministry of Communications, presenting virtually a fait accompli during the initial phases of cellular investment. For the construction of the country's first cellular systems, communities of interest at the local level and in the military had enough authority to shape the introduction of new services.

The Moscow and Leningrad city governments pushed forward with their own licensing, allowing for the construction of AMPS systems over the direct objection of the Federal government. The military subcontractor referred to in the quote, Vimple Communications, soon launched its own cellular system in Moscow using the AMPS. The rollout of these systems was directly facilitated by the local telephone companies, who provided both the capital and some of the critical infrastructure, and the local governments, which passed the requisite laws to assert local authority over these licensing decisions (Law, 1995; Neuman et al., 1995).

Although it would be difficult to prove that companies that focused their negotiations and investment at the local levels were successful than those at the level of the Soviet government, the turn of events during the days of August, 1991, certainly tipped the balance in their favor. The failed coup against Mikhail Gorbachev led to the demolition of the former Soviet State during the final months of the year. The resulting decentralization of authority within the Soviet Union further weakened the hold of the central state on telecommunications laws and regulations.

The Russian Ministry of Communications inherited the weaknesses of its Soviet predecessor. Vladimir Bulgak was chosen as the Minister of Communications, which was split into two portions. The Ministry of Post and Telecommunications continued to have control over the transmission, and Telekom was created to manage the production and sale of telecommunications equipment. This was the first step in the break-up of the Russian telecommunications sector, a process greatly accelerated by the privatization process that began soon after.

The Fragmentation of the Russian Telecommunications System

In 1993, Russia embarked on one of the most comprehensive and dramatic privatization initiatives in recent history. Instead of selling state-owned industry to foreign investors and retaining a stake in the industries, the government chose instead to follow a pattern established by other Eastern European countries making the transition away from a socialist economy. Vouchers were distributed to the Russian people, and the vouchers were converted into shares of companies being privatized. The stated goal of the privatization program was to redistribute ownership to the Russian people and use the power of shareholder relationships to spur new investment and innovations in production (Frydman and Rapaczynski, 1994).

Russia's voucher-led privatization certainly has transformed the Russian economic environment, but the effect has been uneven. For smaller industries and shops, the voucher privatization has been successful in transferring ownership to the workers and managers of small companies located mostly in the urban centers. But the privatization of the larger state-run industries has helped to establish communities of interest that might fight against further reform and liberalization. For the telecommunications industry in particular, this arrangement threatens to stifle investment and undermine the potential for applying new communications technologies to meet the goals of national development.

The "Sharization" of Russia's Private Sector

In larger companies, privatization has been mostly "sharization," with shares being collected by managers, large investment house,s and government entities during the process. Following the patterns of interaction during the communist period, most of the managers have close ties to the political system at the regional and local level. In some cases, the shares have given outside owners the opportunity to oust older managers and institute new forms of management. More often than not, privatization has given existing managers the opportunity to hold on to their positions and maintain their monopoly advantages, even though the long-term viability of such a position is unsustainable; some of those who have purchased shares and do not exercise a controlling interest have even found themselves "frozen out" by managers who write off ownership stakes or dilute the value of the shares owned by those outside the management ranks.

The telecommunications industry is much too large and tied into the existing political infrastructure to be immune from such difficulties. The close relationship between political and industry managers is reinforced by the money-making potential of telecommunications services and the politics of any communication infrastructure. If competition is to flourish in the Russian telecommunications sector, it must begin by a redefinition of ownership and the goals of investment at the grass roots level.

The history of privatization in telecommunications bears the point out. Rostelekom was first spun off from the Ministry of Communications and then broken into operating units according to local geographies. Each of these local telephone companies was privatized separately from Rostelekom, which eventually became the established monopoly provider of long-distance services in the country. The Ministry of Communications officially holds shares in both Rostelekom and the individual, local telephone companies, and governments at the oblast and city level also own shares of the local telephone companies. The problem is that no one is quite sure who owns what percentage of which company.

This, from the business perspective, is the main reason why STET of Italy failed to reach an agreement with the Ministry of Communications in early 1996 on the sale of the various local telephone companies. Although the federal government claimed to hold a share of each of the local telephone companies, exact share percentages could not be defined and valuated to reach an appropriate purchase price. At the beginning of 1997, the Russian goverment changed its approach and folded the local and long distance telephone companies back together again in an attempt to make the privatization deal more attractive to potential investors.

Additionally, a number of "investors without capital" have also played a central role in the privatization process, mostly through their ability to leverage local knowledge of the telecommunications industry and infrastructure. For example, there are few, if any, published figures on infrastructure investment, and no central depository of knowledge as to what and where the actual telecommunications infrastructure in Russia is, though a World Bank study is expected to be published in 1997 with more complete data on Russia's telecommunications development. As a result, people who have that knowledge are able to offer access to strategic portions of a network, such as a switch in one city or a long-distance copper line between two cities that no one else is aware of and no one is using. The quid pro quo that comes in exchange for such knowledge is a portion of the returns from the sharization process.

Because of this weakly defined ownership structure, managers have been able to exercise a wide latitude in establishing services and using revenues. The telecommunications sector has begun to suffer as tomorrow's investment in Russia's future is sacrificed at the altar of today's profit potential. Few speak of a broadly-based consumer market for telecommunications services. Every company concentrates on applying existing models of telecommunications services to skim the cream from the telecommunications market.

The Legacy of the Soviet and Russian Past

The present of telecommunications in Russia is wrapped into the continuing political and economic transformation of the country. Since the break-up of the Soviet Union, the former government monopoly control of all telecommunications services and equipment production has been transformed by a frenzy of local and regional activity. On the surface, the government is still the monopoly provider of telephone services. But, below the surface, a very different set of marketplace conditions has evolved. The monopoly has been undermined by an incredible growth in bypass arrangements. It would be impossible to quantify the extent of the growth or investment in such networks since the partial privatization of the telecommunications infrastructure, but more and more large institutions openly admit that they use other providers for local, long-distance, and data transmission.

Many of these new providers are licensed by the Ministry of Communications, and include companies that are providing dedicated services to banks and other intensive users of telecommunications services. In addition, there are a number of unlicensed providers, including the Russian military and various government ministries, who have cannibalized older telecommunications infrastructures built during the communist period to offer both basic and specialized services.

The official monopoly is certainly an illusion. But to say that true competition is evolving in Russia would also be a mistake. Bypass arrangements can not form the basis of competition for the telecommunications sector because they are constructed to have a narrow focus: Certain people are to have access to certain services. Interconnection is not necessary and, for that reason, infrastructures have been built in such as fashion as to make the economics and technology of interconnection difficult.

There are four different kinds of networks operating in Russia today:

- The Local Telephone Network;

- Overlay Networks;

- Departmental Networks; and,

- Wireless Networks.

Local Telephone Network (Public Switched Telephone Network)

As a part of the privatization program for the telecommunications industry, local telephone service was separated from long-distance services. In the process, Rostelekom was able to shed the older infrastructure of the Soviet era and hand it over to the administration of municipalities and newly created local telephone companies. Many of these companies have become the focal point for local investment. In Moscow, for example, Moscow City Telephone has taken an active role in investing in cellular and bypass networks. Although most of the infrastructure for the local companies is hopelessly outdated, innovative technological patchworks have developed that allow them to graft modern services onto the old, Soviet style network. Cellular services, in particular, rely on a close cooperation between the local telephone companies and the cellular service providers.

Because the regulatory constraints are much weaker in Russia than in the U.S. or the U.K., comparing the local telephone networks with those serviced by the Regional Bell Operating Companies and BT would be a misnomer. The local telephone network in Russia is based more on local political and economic needs than the provision of access to the "public" as a whole. In theory, though, the local telephone networks are supposed to operate in the same way by providing access to customers on the user end and other service providers at the switch level.

Overlay Networks

The term *overlay network* is used here to designate efforts by service providers to completely bypass the public switched telephone network through the use of fiber optic or digital microwave technology. The structure of the digital overlay network is to create a universe of fixed access points, usually in an urban center or among various institutions that have a need for intensive users of telecommunications service, such as banks and large industries. By linking those access points, the service provider can ensure that the companies or people connected to the network receive above standard service, which comes at a higher price.

Overlay networks first appear in Russia in the early 1980s, when companies like IDB and Sovintel began to work with the Soviet government to improve the telecommunications infrastructure of the country. Most Russian overlay networks are hybrid, containing both analog and digital components. The majority are configured with digital switching points and analog connection to subscribers. Wholly digital overlay networks, most notably the digital overlay network run by the Andrew Corporation, employ a combination of fiber optical transmission and international gateway facilities. For Andrew, the strategy was to build sections of its fiber optic network in the metro tunnels of major cities and along the entire Oktyabrskaya railway—a plan that required some close connections with local political and economic interests.

The best analogy to overlay networks in the U.S. is the competitive access providers, such as Teleport and MFS, that have entered into regional and urban markets to target high capacity, business users. But even though the business strategy might be similar, the infrastructure development strategy varies widely from provider to provider. Some are using wireless networks and VSAT terminals to link office to office, others are laying fiber where possible—the strategies diverge widely according to what is available and what can be done with the resources at hand. In that regard, managers of telecommunications concerns have to be much more inventive in Russia to implement this kind of strategy.

But for that very reason, overlay networks are not a panacea to all communications problems. Many overlay networks have not integrated their services with the public switched telephone network at the local level, nor with other overlay networks attempting to provide services. As a result, many of those who subscribe to an overlay network have to use international phone calls to get access to a local line so that they can make a phone call down the street. This kind of arrangement is not only inefficient, but undermines the long-term viability of the local telephone network.

Departmental Networks

During the Soviet period, priority industries and administrative agencies invested in their own telecommunications networks. Because these institutions had priority access to financial resources, they were able to build networks that are technically superior to the public network. Departmental networks were built with technology roughly equivalent to the best in the developed world. The ministry of defense telecommunications system, for example, was completely digital and employed Nokia DX 200, and the Ministry of Foreign Affairs used Alcatel S12 switches. There is no analogy to this in either the U.S. or the U.K.—it is the equivalent of turning over the Department or Ministry of Defence communications system to the private sector.

Today, many of those who know about and understand the whereabouts of the infrastructure of these networks are presently working to revitalize them and employ them for business use. Although many of these projects have surfaced on paper, very few have been launched and even fewer have met their business objectives.

Cellular/Wireless Networks

The Russian government has offered licenses to provide wireless communications services, such as paging and cellular telephone, in a number of the largest urban centers throughout the country. These networks represent the bulk of direct foreign investment in Russian telecommunications. Almost all of the cellular systems involve joint ventures, investments, or part ownership by foreign interests, the most significant example of which is the investments of U.S. West.

Penetration rates for these services are very low, with fewer than 20,000 official subscribers in Moscow. Service providers have chosen to use a traditional marketing and pricing strategy, which only a few can afford, instead of attempting to broaden the usefulness and access to wireless services by working with lower cost applications to push penetration higher. Many purchase the service, not to bypass the network, but rather for the prestige of having a cellular phone. In addition, there are a number of close relationships between cellular providers and the public networks, and some cellular networks reportedly do not have to pay for access to the public network. This kind of business activity will undermine competition if such relationships are allowed to create asymmetrical access standards and costs.

Rostelekom and Its Competition

In one regard, Rostelekom has been a part of each of these networks, insofar as Rostelekom facilities are required to provide services. But as bypass arrangements become more the norm, those links will become weaker and weaker. Additionally, each of these networks has focused on constructing its own infrastructure and its own platform for providing services. It can therefore be said that most of the providers are attempting to position themselves as smaller monopolies in a thin market.

From its establishment in 1993 to its officially planned demise in 1997, Rostelekom focused mainly on creating partnerships with foreign firms willing to invest the capital and knowledge in advanced long-distance and international networks. On paper, the company was organized into a single main center for long-distance communications management, and 17 regional centers. It also provided international access, with an initial near monopoly falling to the company by default until other providers rushed in to tap into this lucrative market.

In total, there are more than 1000 private communications firms operating under Ministry of Communications licenses in Russia, and probably hundreds more that operate without licenses (U.S. Department of Commerce, *Moscow Embassy*

Report, May 1996). Companies competing directly against Rostelekom in the area of long-distance and overlay networks include venerable telecommunications industry names such as Cable and Wireless, along with lesser known companies like Andrew Corporation, which invested more than 100 million dollars to partner with the Oktyabrskaya Railroad Company and the High Speed Railways joint venture stock company to form Rascom, and the Moscow Metro Company to form Macomnet. BT has helped to fund SPI, an international satellite, which competes against Rostelekom and others like Combelga, a joint venture between Comsat and Belgacom, the formerly state owned and operated Belgian PTO.

Each major city has a number of overlay networks competing directly against the local telephone companies, still state owned and still potential sources for political patronage and influence. In St. Petersburg and Moscow alone, the list of companies includes names like Peterstar, BCL, Comincom, Lenfincom, Sprint, Relcom, Sovamteleport.

Names and players change on a daily basis; any attempt to provide a detailed, up to date analysis in a book format like this would be foolish. But, given the history of telecommunications investment in Russia, the pattern of political jockeying at all levels, the evident dominance of cream-skimming business strategies, the disconnection between the market environment and the goals of sustainable development are clear enough. The lack of pervasive, competing infrastructures based on local access and penetration for all customers will undermine telecommunications development in Russia. Together, these networks provide adequate service to pockets subscribers, but these primarily independent networks do not constitute an feasible foundation for an advanced industrial economy operating on market principles.

The problem is not only in service provision, it also appears in areas of equipment standardization and transmission protocols. Within each of these four broad categories, there are literally hundreds of service providers employing the strategy to serve its customer base. Each is using a different type of equipment. Some are analog, some digital; some use the most modern equipment from Europe and America, some use Russian or Eastern European equipment. There is no common technological or customer base to link any of the networks together.

At first, this might not appear to be a problem. In the U.S., for instance, numerous voices have expressed concerns at the balkanization of standards and network infrastructures in the development of advanced telecommunications services. ISDN was a response to that concern, the creation of a standard that was meant to carry higher bandwidth applications to business and residential customers. But creating a standard did not help market penetration in the U.S., or anywhere else in the world for that matter. The innovation had to come from computer makers, developing servers to provide information over distributed networks, and hardware providers, creating cheap models to allow individual computers access to the servers on the existing copper-wire telephone infrastructure.

Unfortunately for Russia, it cannot rely on such a development pattern. Developed countries, because of the political control structure placed either directly or indirectly on the activities of telecommunications providers, have already put a distributed network in place based on relatively homogeneous and consistent standards. They also have the resources available to waste money and time on experimenting and theorizing before creating a more optimal solution. Even in developing countries, the future of telecommunications is in distributed networks, not isolated networks. Service providers have to offer access to any and all applications and information available, while simultaneously using all of the portions of the network to efficiently distribute the volume of information being sent through the network. If networks do not have a common technological platform or economic framework to connect with each other, the creation of a unified, distributed network is not possible. That severely hampers the ability of service providers and subscribers to gain access to all services available.

The result is economic inefficiency. For example, subscribers to the overlay networks often have to make an international call to get a local telephone line. Today, this may seem only an inconvenience, but when demand for applications such as electronic data interchange (EDI) increase, and the required interconnection arrangements cannot be insured, the market for such services will fail to materialize. Without EDI, the Russian financial community will be severely restricted in its ability to grow and profit from the recent advances in telecommunications services and infrastructure.

Sustainability in the Russian context, in that regard, has less to do with merely implementing an infrastructure solution than it has to do with shoring up the very foundations for legitimacy and authority in the governing structures. The problem, though, is that history appears to discount the chances of countries with weak political systems successfully implementing telecommunications reform (Petrazzini, 1995). And, as discussed in earlier chapters, political legitimacy depends, in great part, on good policy and political leadership, and, although good policy ideas may be in circulation, Boris Yeltsin's deteriorating health would indicate that a protracted silence from the most senior government official on all difficult issues may be likely.

So there is both service provision and political legitimacy to consider. To improve service to existing consumers of the Russian telecommunications industry, and to increase the availability of low-cost, high-quality service to more consumers, both of those issues will need to be addressed. Before a specific discussion of the regulatory and business strategies in place in Russia's quickly expanding wireless services market, it is worth reviewing, in this context, some of the barriers and constraints which exist in implementing a liberalization and full privatization program in Russia.

OVERCOMING BARRIERS TO LIBERALIZATION AND PRIVATIZATION POLICY IN RUSSIA

The Russian case has unique qualities that differentiate it from other developing world contexts; the transition from totalitarianism, combined with size and historical roots of the country, produce a complex pattern of political and economic relationships. The preceding discussion has pointed to the importance of local and regional telecommunications development in Russia, the conditions for which emerged in the collapse of the Communist system in the early 1990s.

But many of the barriers to implementing liberalization and privatization policies in Russia exist on a national scale; regional groups can continue to invest and expand infrastructure resources through quasi-private and semi-autonomous corporate entities, but the legal and business foundations for this kind of activity are tenuous at best. This section steps back from the discussion of individual network types and strategies to outline four of the most critical barriers and constraints preventing the implementation of privatization and liberalization policies, including the monopoly characteristics particular to the Russian case, licensing policies, spectrum management, and financing.

Monopoly Characteristics Particular to the Russian Case

Any time a new market opens up, there is a tendency for firms entering the market to aspire to monopoly. Each firm assumes that it will do best by controlling the entire market. When the market is large and the number of competitors is relatively small, there are incentives for competitors to divide the market among themselves and agree not to compete with each other. Whether by design or by happenstance, this type of anti-competitive collusion appears to be occurring in Russia, and the present organization of the Russian telecommunications industry reinforces monopoly over competition. This can be seen in the three kinds of monopolies that have developed in Russia since the fall of the Communist regime (Neuman et. al., 1995): monopolies established by geographic area, by closed access, and by service provision.

The aspiration to monopolize services in a given geographical area is more than a business proposition—monopoly is often good politics, especially for communities that are used to monopolies based on state control. Each region and locality has established its own "monopoly" service provider for local telecommunications, and some of the local authorities (Moscow included) have passed laws that reserve to the local government the right to renationalize the telecommunications provider if the community's interests are somehow threatened (Law, 1995). Outside of Moscow, there is a movement toward local authorities establishing exclusive franchises for overlay and data communications services.

The difficulties with these monopolies are both administrative and technological. Relationships with different localities and carriers vary, requiring new negotiations at each level for any company wishing to do business in Russia and gain access to telecommunications services. Additionally, the quality of the infrastructure varies widely from region to region, sometimes requiring different technological solutions for interconnection between the individual monopolies. At best, this causes economic inefficiencies and increased costs; at worst, it can cause a business venture to fail.

Quite often, service providers simply refuses a request for interconnection from another carrier or company wishing to have access to telecommunications services. This situation can be though of as creating a monopoly by closing access to services. Because there is no open regulatory design that has been developed by Russia to mandate interconnection between networks, such conflicts are usually dragged through the political process in a individualized and personalized fashion; no attempt to create a broad standard for interconnection has yet been attempted by the legislature or Ministry of Communications.

In the U.S. and U.K., the obvious remedy for this situation has been to rely on either negotiation or the dictates of the regulator to solve the problem of interconnection. With the passage of the 1996 Telecommunications law in the U.S., numerous interconnection agreements between local telephone monopolies and other service providers were signed and announced—no such flood has washed through the telecommunications markets of Russia. That is in great part because the Russian providers have no incentive to negotiate such agreements, while, in the U.S., monopoly service providers wishing to enter new markets have an incentive to bend. Without a strong regulator, another incentive needs to be communicated through the market that will drive changes in corporate behavior.

Then there are monopolies by service provision. In major urban centers, for instance, there is a division of digital services by bit rate, the established measure for the volume of information that can be transmitted through a telecommunications network. Various carriers have staked out communications links with a particular speed, such as the standard phone line for voice (1-2 kilobits per second), voice network modems (9600 bits per second), and various higher speeds (most predominantly at 64 kilobits per second and at 2 megabits per second). The difficulty is that, in fixing a bit rate, service providers often make it impossible for interconnection to occur with a transmission system at a lower or higher bit rate. The opportunity for "bandwidth on demand," which would allow users to flexibly choose the rate of data transmission to meet their needs, does not exist in practice.

These monopolies have built to achieve competitive advantage in niche markets or according to the regulated monopoly license provided by a local or the national government. Needless to say, there is nothing unified or open about this arrangement: Access is closed, and arrangements that bring networks together are almost non existent.

The monopoly strategy makes sense for a single firm, but not for the society as a whole. The Ministry of Communications and the local authorities must resist the temptation to conspire in the establishment of little monopolies. Where there is an economic advantage to keeping competitors out, there are incentives to persuade the gatekeeper to close the door on later competitors. That can not be allowed to happen. If Russia is to establish a unified and open communications environment, it cannot be through the establishment and cultivation of monopolies. Such an environment requires competition and open access.

Constraints to Competition: Licensing Policies

The goal of promoting competition in the Russian telecommunications industry will be attained much more rapidly through the adoption of an effective licensing model. One common conception of licensing is the sale of exclusive franchises, and many of the licenses granted by the federal government employ this model. This kind of license establishes a monopoly, and there are obvious incentives for corruption when monopolies can be arbitrarily created through licensing. The impact for the society as a whole is also questionable; the incentives monopolists face lead to small scale investments in the most profitable regions and technologies, and monopolists have no economic incentives to lower prices or service smaller customers.

The goal of licensing should not be to limit the number of firms operating in the market by creating monopolies. Rather, licensing should serve the purpose of insuring adherence to the ground rules of the Russian telecommunications regulatory authorities. The ground rules should be limited to technical standards and interconnection to competitors, but also might include proscriptions for the building of new infrastructure and the percentage of customers served over time.

Licensing is a means of enforcing regulations, and the value of a license is should be proportional to the amount of resources the licensee is willing to invest. The result of open access through low-cost licensing is greater levels of investment, more competition among service providers, and lower cost service to a larger number of consumers.

At the same time, licenses should be flexible enough to allow service providers in one area to begin competing with licensees providing other, similar services. A provider of wireless services is a facilities-based competitor with wireline local service providers. The ability to offer "generic" wireless access on designated spectrum offers the most reasonable and open system for ensuring that facilities-based competition can emerge. But, by constraining cellular providers to specific technologies and specific kinds of service models, the government has effectively drawn a line that prevents the entry of a wireless access provider into the market for local services.

An open license policy that proactively helps shape the market serves to direct the appropriate signals to service providers over time. But licensing is only part of the equation. Obtaining a license to compete against a dominant firm is only part of the battle to enter a market—without interconnection to the dominant service provider's network, new firms will fail to attract customers.

Spectrum Management

Before the break-up of the Soviet Union, there were no market incentives for efficient spectrum use. Because there was no competition for spectrum space with commercial users, government and military users were not rewarded for using bandwidth-conserving technologies. Today a large invested stock in spectrum-inefficient equipment impedes the transfer of spectrum to other users. At a time when there could be profitable use of the spectrum for civil and commercial uses, the military is refusing to relinquish spectrum. It would indeed be costly for government and military authorities to upgrade their equipment to more efficient technologies, but the management of spectrum could be arranged to satisfy both current and new services.

For the government agencies and military to see an incentive for relinquishing spectrum, they must be offered the opportunity to benefit from the transfer to other segments of the spectrum. The possibilities for this are numerous, but two principle incentives are available. First, spectrum auctions could be used for allocations of new commercial services. The proceeds from auctions would go to reimburse current spectrum users who would have to invest in new equipment to make way for new commercial services. Second, the military and its industrial affiliates could benefit by manufacturing equipment for use in new commercial services. The military industry's experience in the manufacture of radio equipment could be exploited in the marketing of new wireless local loop equipment and the installation and maintenance of all sorts of communications and computational equipment. Finding commercial uses for Russia's know-how will prevent brain drain, both to foreign countries and to commercial concerns within Russia. It will also serve to maintain a sufficient and efficient industrial base to satisfy the technical needs of the Russian military in the future.

The combination of licensing, interconnection, and spectrum management will give the government an opportunity to establish market mechanisms that will alter business practice. In combination, these kinds of regulations will establish a unified structure for telecommunications services, where data and traffic flows move uniformly through different service providers' networks and market entrants are given the opportunity to promote new products and services.

Financing: The Critical Barrier

Clarifying the rules of the economic game in the telecommunications sector, though, is only a first step in creating the foundation for the development of the telecommunications sector in Russia. Improving the investment environment and proactively bringing investment into the telecommunications sector are two different things. Considering the low quality and penetration rates for telecommunications services in Russia, it is clear that investment in new technology will be critical. To achieve a 35% teledensity, Russia will need at least 34 million new telephone lines at a cost of approximately $82.6 billion.

Separate projections by the Organization for Economic Cooperation and Development (OECD) and the International Finance Corporation (IFC) indicate that 40% to 60% of this financing for new investment should come from operator revenues. Direct foreign investment should account for approximately 25% to 35%, with a combination of municipal debt and equity financing taking on the rest. How can Russia get the money to build the communications networks it needs?

To date, Russia has concentrated mostly on the local generation of debt and equity financing through the privatization process. Foreign players have not been brought in to invest in and manage the wireline telecommunications infrastructure. Although some claim that the necessary investment for rebuilding the Russian telecommunications infrastructure can be generated from within, the present arrangement presents a number of difficulties.

First of all, foreign direct investment has been limited to specific projects and has not had the broad-based impact that would help push penetration rates higher. According to the United Nations Economic Commission for Europe, Russia has received only $36.5 billion dollars in long-term commitments across all sectors of the economy, a number that is less than the total for Kazakhstan (*Wall Street Journal*, April 12, 1995). This is a fraction of what is needed for the telecommunications sector alone.

The only area where foreign investment has played a direct role in the construction of local telecommunications services has been with emerging communications technologies, such as cellular communications. Projects have been financed largely by foreign providers in partnership with Russian companies. But the sums required to set up a small cellular system to support only 15,000 users is a trifling sum when compared to the investment requirements for the Russian information infrastructure as a whole.

Investments in long-distance service, exemplified by the 50 x 50 project, also have done nothing to facilitate the penetration of local telephone services required to construct a consumer market for telecommunications services. The name comes from the extent of infrastructure construction that the project was meant to fuel: 50,000 kilometres of fiber optic cable to connect 50 cities throughout Russia. The principal players were U.S. West, Deutches Bundespost Telekom and France Telecom. After 2 years of negotiation, the project was launched in 1994, even though there was no agreement at the time between the partners about their contributions or share of revenue.

The 50 x 50 project succeeded in financing the creation of three new international fiber optic cables stretching into Europe and Asia, but the promise of increased local penetration through reinvestment appears to have been an illusion. No concrete plans for further investment in the project and shared revenues from the project have appeared, and it would seem that none are forthcoming. As a model for increasing investment in the country, 50 x 50 also misses the point: It has given providers an excuse to invest in long distance and international linkages while ignoring the needs of consumers at the local level.

Investment levels remain low even in these long distance and specialized networks that would be most useful to intensive users of telecommunications services. One professional estimates that only $20 million has been invested in data networks throughout Russia since 1993, a sum that pales in comparison to the amounts invested by banks and other institutions throughout the developing world during the same period of time (Neuman et al., 1995). It would appear that specialized networks for data transmission are not likely to fuel broad-based investment in the country's telecommunications infrastructure.

Local financing, on the other hand, can be an extremely difficult exercise in a tenable economic and regulatory environment. Telecommunications is, by its very nature, a capital intensive industry, requiring the investment of hard currency and the use of foreign reserves. During periods of inflation, local prices tend to lag behind the flux in the currency, making it more difficult for local operators to use revenues to fund continuing infrastructure investments. The option of floating debt denominated in foreign currency, and thereby retaining ownership while attracting new capital for investment, runs a double risk in this environment: The company would no longer be able to generate revenues to repay the debt, and investors might find the purchase of debt less than attractive when compared to the potential upside of equity financing offered by other countries and companies.

Additionally, Russia does not have a highly developed and mature enough credit market to sustain a strategy of predominantly local financing, which is a critical part of ensuring sufficient investment in infrastructure resources. If the financial community is not yet prepared to efficiently allocate resources to a broad based investment program in infrastructure resources, it is necessary to rely on foreign investment for a greater percentage of the financing.

An example from abroad is instructive in this regard. The World Bank, in its 1994 World Development Report, compares the sources of funding for infrastructure investment in East Asia and Latin American, taken as regions. East Asia has a strong and deep credit market, anchored by Hong Kong, Japan, Singapore, and the local exchanges of the growing countries throughout the region. More than 90% of infrastructure investments have been funded by internal capital sources. In Latin America, which historically is dominated by large, state-owned and operated banks, more than half of infrastructure investments have come from abroad (World Bank, 1994).

Russia has the opportunity to steer the middle road between these two extremes, balancing the evident strength of the emerging banking community with the directed resources of interested foreign investors. But to chart that middle course, Russia needs to identify strategic opportunities to push penetration rates higher and improve the possibilities for investment to have a real impact on increasing the capacity and quality of the Russian information infrastructure. These opportunities will allow Russia to leapfrog past the existing technologies that dominate the provision of telecommunications services in developed countries (Kazachkov, Knight, and Regli, 1996).

Simultaneously, they would allow local interests to continue their strong investment and interest in local telecommunications services. This is clearly a priority for the Russian government, since the telecommunications sector has been defined as a "strategic asset" for the country—there are no telecommunications companies operating in Russia which are 100% owned by foreign entities, and the Ministry of Communications reports that the average foreign share of Russian telephone companies in the 86 regions of Russia does not exceed 5.2% (U.S. Department of Commerce, 1996).

What is clear in the discussion of these barriers is that good policy will come at a premium in Russia. Much of what has been outlined in these four subsections would be a solid agenda for telecommunications reform in a country with the political will and ability to act on the precepts, but, in a country where virtually no coordinated action is possible in charting a course for telecommunications development, any one of these barriers is enough to choke off investment and future growth.

In that regard, wireless access has a critical role to play in the development of Russia's telecommunications sector—not just because it can provide access, or lower costs, or better, more reliable service in areas where the existing telecommunications network barely reaches. A strategic focus on wireless access may aid Russia in a more fundamental way, allowing the success of investment and good public policy to build legitimacy for national and local government in an area critical for the future economic and political development of the country.

The previous sections have outlined some of the history of wireless communications in Russia, as well as part of its role in the present structure of the country's telecommunications networks. The next section takes that discussion one step further and begins to compare the possibilities of the infrastructure with the specific challenges already laid out.

THE OPPORTUNITIES FOR
WIRELESS ACCESS IN RUSSIA

At the end of 1996, the Ministry of Communications estimated there were about 160,000 mobile subscribers in all of Russia. Considering the fact that, at the same time, there were almost as many subscribers in nearby Estonia (a country with less than 2 million people) than in Moscow (a city with more than 9 million), that number is not representative of a great accomplishment. (U.S. Department of Commerce, *U.S. Embassy Report*, February 1997). Even so, estimates are for a 300% increase in the number of subscribers in 1997 alone—and that perhaps more than a million people will be subscribers by the year 2003.

Part of the problem is clearly the prices being charged: Set up costs are reported to be as high as US$2,000, with 65-70 cents a minute charged for airtime. The present pricing and business strategy for wireless access services in Russia is a direct reflection of the cellular model that has been rolled out in the United States and the developed world over the past decade (Law, 1995). In a country with a significantly lower per capita income than the U.S. or the U.K., these prices are well outside the range of the typical Russian consumer.

But that is not really the target market for the companies pioneering the opportunities for wireless access in Russia. They are looking to acquire the businesslike and wealthy of the major urban centers, offering cellular as a bypass solution to the wireline network.

The history of corporate involvement in the Russian wireless market bears this fact out. As of April 1996, there were 196 registered providers of cellular and mobile communications in Russia. Some of the most important names, like U.S. West and Millicom, have already been mentioned.

By the end of 1996, local governments had largely distributed the portions of the spectrum that became their responsibility by default: 800 to 860 MHz, opening the door for the construction of systems similar to the analog cellular systems operated in the U.S. The Federal government, after the enactment of the 1994 regulations adopted by the State Committee on Telecommunications, began the distribution of licenses for GSM systems in the 900 MHz band for a variety of locations throughout the country. Some of the licenses offered by the government were apparently not of interest to various bidders as no offers were tendered, but many of the urban centers now technically have licensees in the 450 MHz, 800 MHz and 900 MHz bands.

Reports are that a few Russian companies have the inside track to receive licenses for PCS services in the 1800 MHz range, including Vimple Communications and Rosiko. These licenses would allow the companies to expand their existing service networks in Moscow and possibly complete the process of expanding to national coverage for services.

For a company like U.S. West, that has required the management of a number of different relationships throughout the country. The company's portfolio of holdings in Russia, all managed by the Russian Telecommunications Development Corporation (RTDC), include 450 MHz licenses in Moscow and St. Petersburg (Moscow Cellular and Delta Telecom), GSM service in the 900 MHZ range in Rostov-on-Don and Nishni Novgorod (Dontelecom and United Telecom), NMT standard service in the 450 MHz range in the Ural and Siberian regions (Uralwestcom, Baikalwestcom, Yeniseytelecom), and 800 MHz AMPS service in Vladivostok (AKOS).

Commentators who feel that the PCS licensing process has balkanized the U.S. wireless services market need only look to Russia to see that there are far more perplexing arrangements possible. The U.K. environment, with its four existing providers and range of providers in waiting, the advantage of a smaller market is evident; the reach across 11 time zones and such a varied geography presents a management challenge that may just be the most significant available for a telecommunications company operating in a global marketplace.

Even in a confined market like St. Petersburg, the difficulties in cobbling together individual networks make any theoretical discussion of how to promote effective competition for wireless services appear irrelevant. As of mid-1996, three wireless service operators provide cellular-type services in the St. Petersburg region: Delta Telecom, North-West GSM, and SPT (St. Petersburg Telecom). Delta, the first provider, operates at 450 MHz with the NMT standard, and SPT runs an AMPS system provided by Motorola. Both of those licenses, as previously noted, were offered by the local government and effectively "legitimized" retroactively by the Ministry of Communications. Delta's coverage range is extensive, but that is in great part because the lower frequency license allows for more coverage with fewer cell sites; the company claims to serve more than 16,000 kilometres in and around St. Petersburg. In total, Delta is close to having 20,000 subscribers officially on its network, while SPT only has 5,000.

North-West GSM operates a system at 900 MHz that was licensed by the Federal Government. By the end of 1995, it had 2,800 subscribers, but was predicting growth to more than 16,000 by the end of 1996 (U.S. Department of Commerce, 1996). It has focused largely on the urban center, offering the traditional cellular systems and focusing almost exclusively on higher cost, mobility based applications.

A whole range of paging firms operate in the major cities, but none have subscriber bases larger than a few thousand. In St. Petersburg, for example, Neda-Paging claims more than 6,500 subscribers. But there appears to be little room made by any of these providers to go beyond the basic parameters set in the developed world; cellular type systems are the mainstay of wireless access provision in Russia.

The opportunities for wireless access in Russia are therefore constrained by the kinds of services presently provided. The difficulties in establishing a legitimate and effective pattern of telecommunications development in the country has constrained the space that is open in a competitive marketplace for business innovation and new kinds of investment. If Russia is to move beyond the cellular model, what are some of the directions that the country should consider and how do those directions help to equip managers in the public and private sector with the knowledge and resources necessary to establish a new model for telecommunications development in Russia?

Wireless in Russia: Breaking Through the Foreign Model

Early investments in cellular systems in the U.S. and U.K. were provided on the backs of highly leveraged balance sheets. This strategy, in many ways, was transferred to the developing world and is reflected in the prices; prices are targeted to increase the cash flow so that profits, when possible, can be repatriated to the countries and companies that are the source of the investment. The pressure to increase cash flow comes not from junk bonds, as was the case for McCaw Cellular in the U.S., but from foreign investors, whose assessment of the levels of risk require the expatriation of additional cash so that returns on investment are ensured.

Nevertheless, in Russia, there are conflicting reports about the profitability of the investment and the ability of certain companies to extract the revenue from wireless access operations. Specifically, it is estimated by some that the net annual revenue from the U.S. West/MCC/Milicom cellular license in Moscow is more than the total investment in infrastructure made by the participating companies, but that the Western partners have not seen the expected payment of returns to date.

The impact of that strategy can be seen in some of the preliminary customer research that has come out of Russia. Preliminary research from one of the cellular carriers in St. Petersburg indicates that the peak time for transmission on the system is from midnight to 2 a.m., indicating that certain kinds of business appropriate to that time of the evening have been driving usage and revenues for wireless services. This is certainly not the replacement of the landline telephone and the increase of competition of which wireless access is capable.

If investors have been unable to achieve a return on their investment, and the long-term goal of increased penetration can not be achieved through the implementation of the cellular model in Russia, a better course has to be charted. Part of that better course is to identify local funding opportunities that go beyond the traditional joint partnerships where the foreign firm provides the finance and the local firm provides the access. Russia, like many developing world countries in the midst of economic transition, has reserves of wealth abroad that were gathered during the first few years of capitalist accumulation. Cyprus is generally regarded as the destination of choice for legitimate and illegitimate business ventures to take cash offshore and place in storage until the business climate becomes more stable.

What may drive local investors to repatriate cash is the establishment of a truly Russian solution to Russian problems—technology developed in Russia for export to foreign countries, systems and models for access and provision that can be shared in successful business ventures abroad. This is a vision expressed often in the past telecommunications policy of a number of countries, and it was behind Brazil's efforts to develop a local telecommunications technology presence for itself since the early 1970s. In Brazil's case, that investment has brought a relative degree of success, leaving a legacy of untapped potential that could reappear if the restructuring of the country's telecommunications market ever does take place. From a financial point of view, though, those are the kinds of solutions that have made returns for local investors and, at the end of the day, investment is not going to occur unless the business climate provides an opportunity for return commensurate with risk.

That is why this kind of model would be best expressed through government institutions. These are the institutions capable of at least setting the terms of the debate, even if the level of legitimacy required to enact and enforce regulation may not exist. Given the relative youth of the wireless services market, and the lack of variation in the models being implemented, it represents the best opportunity to articulate a oherent vision of a Russian model for telecommunications.

The Economics and Business Policies for Wireless Access in Russia

What is the nature of that model, then, and how might it be elaborated? At the heart of the redefinition of wireless access in Russia is a reduction in the cost of providing the services. Administrative, sales, and marketing costs can be reduced through the outsourcing of functions of the wireless network, such as billing systems and sales forces for retail sales. Capital costs, which include the cost of switching, the cell site which receives and transmits information back and forth to the subscriber, and communication between the cell and the public network, assume a network functionality that includes mobility. As more networks are constructed to facilitate mobility, the more complicated the system becomes and the more expensive the capital costs are to construct the network. By implementing a wireless access solution that involves fixed access points, it will be possible to bring down the capital costs substantially.

This is the goal of wireless local loop: bringing down costs and increasing local penetration of services through appropriate wireless access technologies. Through the application of this technology, Russia can establish the foundation for a broadly based consumer market, contrary to the arguments of many who claim that no such market can exist in the foreseeable future.

Price is not the only advantage of strategically implementing wireless local loop to facilitate the penetration of services. It is much easier to scale investments in wireless technology to the needs of a specific market or market segment than it is for wireline telecommunications. This factor provides an even stronger business rationale for the strategic implementation of a wireless local loop architecture.

As discussed in the second chapter, an investment in wireline technology is extremely capital intensive. Not only is it necessary to connect each of the individual points on the network with a dedicated line that can stretch from 20 to 100 meters in length, but it is also necessary to put into place a trunking system that can carry all of the traffic from each of the individual subscribers. Calculating demand becomes a very precise science, and a very important one as well. Overbuilding causes financial loss that can not be recovered, and underbuilding can seriously hamper the ability of the network to provide quality service and expand to handle new traffic and subscribers.

Wireless technology is built on a scalable architecture that can be expanded and contracted much more easily to meet shifts in the demand for telecommunications services. For example, if a new subscriber base emerges in a different portion of a city or region that is serviced by wireless access, it is not necessary to rebuild the system and wire the new territory. All that is necessary is the establishment of a new cell site that can support the new subscribers.

For Russia, this kind of flexibility will be critical. As an economy in transition, it is likely that there will be strong increases in demand, often in geographical areas that might not appear to be a high priority for business investment at the present time. For example, as demand increases in highly populated urban areas with a higher demand for telecommunications services, cell division will enhance the capacity of the network. For a wireline network, the only option would be to rebuild the trunk lines and construct new drops from the trunk line to the homes.

Rebuilding trunk and drop lines would be extraordinarily expensive and difficult. First, the sheer lack of capable people needed to complete the task makes the execution of such a strategy almost an impossibility. Further, continuing difficulties on the definition of land rights and titles would make any such work a serious business risk; one day, the line is on a plot of land owned by the company or rented by the company, the next day the land may be in completely different hands.

As a strategic business decision, wireless access is much safer and much more capable of meeting changing demand conditions. In that regard, the appropriate business decision for any new or existing provider is to begin the transition to wireless local loop architectures in the provision of basic telephone services throughout the country.

The Public Policy of Wireless Access in Russia

Some of the critical pieces of the public policy puzzle have already been discussed in the section on barriers to the reform of the telecommunications markets in Russia. Even so, there are two specific public policy issues critical to the establishment of a competitive framework for wireless access services in Russia: spectrum management and interconnection with the public switched telephone network.

Spectrum Reform for Wireless Access

In order to provide wireless local loop services, Russia must address the larger problem of spectrum management. Spectrum management has become bifurcated, with localities pushing the AMPS standard for cellular service at 800 MHz and the federal government licensing GSM. The Russian military still controls large swaths of spectrum and, in some cases, has already begun to use the spectrum under their control for commercial purposes. Russia is in desperate need of a law to help regulate control of and use of the spectrum, but the law needs to be open and flexible enough to ensure that innovative, new services like wireless local loop can be offered. The Duma attempted to reform spectrum control in 1995 and 1996, with the issue scheduled to be raised again in 1997, but resistance from the beneficiaries of the current system have blocked change.

The only answer to this difficulty is for Russia to establish an independent spectrum commission, and empower the commission to monitor the use of the spectrum by service providers and create regulations for licensing and the resolution of conflicts. Regulations should be executed and enforced by the Ministry of Communications, rather than made by the Ministry.

One of the first responsibilities of the independent commission should be to arrange for the auction of licenses in the 1800 MHz to 2200 MHz range, which will be used for the provision of PCS services in most developing countries. Because equipment has been developed for this frequency band, it will be easier for service providers to get access to a number of equipment manufacturers that are prepared to support the construction of wireless infrastructures. The proceeds for this auction should not be expected to be very large, especially considering that no telecommunications company has shown an interest in some of the licenses for the 800 and 900 MHz bands in the more isolated regions of Russia. Nevertheless, it is likely that licenses for Moscow and other large cities would receive a fair value in an auction proceeding, or that national licenses would receive real attention from local and international investors.

Interconnection for Wireless Access Service Provision

The transformation of the telecommunications industry from a wireline to a wireless infrastructure will not happen overnight. For that reason, it will be critical for the Russian government to establish standard practices for the connection of wireless and wireline telecommunications infrastructures. At present, the Russian government in general, and the Ministry of Communications in particular, has done more to hamper the interconnection between network systems than they have done to facilitate it.

The technologies required for the interconnection of wireless systems into the public switched telephone network and other wireline networks have become entangled in the politics of infrastructure investment in Russia. Service providers have been innovative in developing mechanisms for linking together the most modern telecommunications equipment with the older technologies of the Soviet telecommunications infrastructure. The Russian government has chosen to view these innovations as an impediment to their ability to coerce foreign service providers to introduce the most modern communications technologies. The Ministry recently mandated that intermediary standards can not be used to connect cellular services to the public switched telephone network, a decision that contributes to the fragmentation of the market by hampering the ability of wireless access providers to develop mechanisms to quickly increase capacity and meet demand.

The public policy problem is also one of incentives. As we have already mentioned, there are many different kinds of monopolies in Russia, and there are presently no incentives to facilitate the interconnection of networks, be they wireline or wireless systems. The government must adopt a set of "sticks and carrots" to incent companies to connect their networks together over time.

The "sticks" used by developed countries would be appropriate to the Russian case, insofar as they can be enforced by the Ministry of Communications. Legal sanctions against companies who do not open their infrastructure for traffic from other providers will be critical, and clearing the way for appropriate legal challenges will play a critical role in establishing sanctions against recalcitrant providers.

The "carrots" are somewhat harder to identify, but will be more critical in the long term. First, the Ministry must take the public position against the continuation of the restricted monopoly system and provide clear business advantages for the integration and interconnection of networks. Those arguments will be based on economies of scope and the ability of each of these networks to achieve a broader base of customers over time, thus increasing the possibility of improved revenues and infrastructure value. More concretely, the Ministry and members of the legislative branch could advocate specific tax advantages for telecommunications providers that carry the traffic of other providers, thus offering a clear financial argument for interconnection.

STRATEGIC LIBERALIZATION IN RUSSIA

These are the kinds of directions in corporate and government policy that will help Russia overcome the fundamental barriers to sustainable telecommunications development through wireless communications. Throughout the discussion, comparisons to the U.S. and U.K. situations have been made, even though the differences between developed and developing worlds make any comparisons difficult and occasionally untenable.

Nevertheless, it is worth laying out the same three areas of discussion in presenting what the theory of strategic liberalization might mean in a Russian context. How might some of the ideas on the level of government intervention, the substance of government policy, and the direction of corporate policy inform public and private sector managers interested in creating an environment for sustainable telecommunications development in Russia?

The Level of Government Intervention

In the U.S. and U.K. cases, the discussion focused on the need to step back and reduce levels of regulation in the wireless services market. Two specific areas where the level of government intervention needs to be reduced in both of these countries are in license definitions and seamless service provision—unless service providers are able to provide what the market demands, it will be difficult to establish a pattern of evolution and growth in the market for wireless access services.

These two issues have less relevance in an environment like Russia, where the rules that do exist are barely enforced and the rules that should exist are only on the drawing board. By default, companies are providing what they want to provide because the licenses they were granted are often closely connected to local political interests and needs. And the patterns of cross-subsidization between local telephone providers and cellular telephone providers implies that seamless services are already a reality, even though they may be damaging the ability of competing providers to establish a foothold in the higher end of the market.

The regulatory, centrist approach would probably suggest a regulatory solution, calling for separations to be put in place and better divisions to be forced on the existing providers. But the inability of the government to function effectively in providing this kind of guidance and enforcing uniformly the divisions established clearly indicates the ineffectiveness of this approach in the Russian context. Solutions need to come from the ground up, rather than from the top down.

Given the limited amount of resources available to the Russian government, and the continued strong independence of local and regional governments, the best option is to choose areas where policy can be built rather than attempting a wholesale transplantation of the regulatory regime of the developed world. To a certain degree, the tendency will be to retroactively certify the activities of strong interest groups and corporate players as they continue to push their investment programs further, but the opportunity will exist to build areas of consensus from individual actions. In that regard, the two categories presented in the U.S. and U.K. discussions might have a great degree of relevance after all.

License definitions, both in wireless and wireline services, have been defined in terms of technologies and franchise territories, the same as in the U.S. and the U.K. The U.S. and U.K. examples have made it clear, though, that license definitions become part of the problem, simply because they hamstring corporate players who wish to expand the range and type of services available to customers. Russia may soon find that it has classified services retroactively, only to find that a new set of services is evolving and that existing providers have gotten out ahead again in areas like distance learning and distributed applications—even though the language of the licences indicates those kinds of services are not in the company's purview (Kazachkov, Knight, and Regli, 1996).

One option would be to use the common law traditions in the U.K. and expect constant license revisions. But the traditions of the U.K. are probably not exportable to the Russian context, where shifts in legal arrangements are likely to be seen as more ineffective attempts to regulate and restrict the industry's growth and development.

Licenses, especially for wireless services, should therefore be defined not in terms of specific technologies, but rather as generic licenses for the provision of telecommunications services. Seamless provision, already a reality, will have a framework to sustain itself and corporate entities will be able to establish a consistent legal space from which they can elaborate specific kinds of strategies.

The important reality is that there are options other than the regulatory option, even in a developing world context where the government has little ability to affect change in the marketplace. Although this might not be sufficient to counter the claims that weak political systems are less likely to implement successful telecommunications liberalization policies, it at least opens the door to the possibility that policies might not need to be implemented to provide a foundation for sustainable telecommunications development.

The Substance of Government Policy

Within that context, it is clear that the substance of government policy in the Russian case needs to be oriented toward increased innovation through competition, rather than through protection and subsidization. The resources are there, in the

Russian military, in the academic and scientific community, and in the private sector, to finance some truly innovative work in telecommunications technology and applications for use in distributed networks. It may be possible for Russia to capitalize on those assets and use the U.S. model for preference in support for technological innovators.

Preferential Treatment for Technological Innovators

To increase service penetration and decrease the costs of providing telecommunications services, Russia will have to place itself at the forefront of technological innovation by becoming a laboratory for and source of new breakthroughs. Innovators should therefore have preferential treatment with regard to resource and license allocation where possible.

The critical difficulty in this proscription is the identification of the technological innovation and the establishment of an appropriate benefit for the innovator. As the American case has shown, this process is inherently political and can create impressions of favoritism and patronage. Russia will have one advantage, though: the experience of other countries where wireless access has already achieved relatively high rates of penetration. If the Federal Communications Commission, for example, determines that a real technological breakthrough has occurred, it will be possible for Russia's Ministry of Communications to follow the Commission's lead and motivate the innovator to bring the product to Russia. This kind of activity will be critical to Russia's ability to become a early adopter of innovative technology, and can help to depoliticize the activity somewhat. Nevertheless, specific procedures for the review of domestic products and their classification will be critical to fostering the innovation among local producers in Russia's telecommunications sector.

The Direction of Corporate Strategy

The discussions in the U.S. and U.K. sections on the direction for corporate strategy have concentrated on elaborating strategic liberalization as a business proposition, providing the rationale for lower cost, fixed wireless local loop services in a developed world environment, where a greater percentage of customers are willing to pay the price for mobility. Commentators are often quick to point to the developing world, though, as the most likely market for wireless local loop simply because rates of penetration are so low and incumbent providers are less likely to be able to respond in meeting increased demand.

What those commentators fail to point out, however, is that just because the technology *can* respond to market pull, interests quite often push in the opposite direction. Incumbents may not have a 99% penetration rate to protect, but that does not mean they will look at wireless local loop providers as relieving them of a market burden that they could not bear on their own. The political economy needs to be a focus for corporate strategy, which runs against the expectations of the techno-libertarians and others on the right side of the political spectrum who feel technology alone can answer the problem of market failure.

Russia is a perfect example of this. Everywhere a successful cellular provider has flourished, a local telephone company is behind the expansion. Wireless local loop has to be more than a good idea; there have to be incentive structures in place that help companies support low cost services and still maintain reasonable levels of cash flow.

Two areas where discussions in the past two chapters concentrated were decreasing the costs of provision by outsourcing marketing, sales and billing functions and decreasing network maintenance costs by creating incentives for the colocation of facilities. Both of these have relevance to the Russian case and the opportunities for building a strategic liberalization policy.

Outsourcing

Outsourcing represents perhaps the most important option for cost cutting and reducing the cost of services in the developing world—quite often because the price for real managerial talent is prohibitively high. Specialized functions, like engineering resources, billing systems, marketing, and the like is in very short supply and, because of the laws of scarcity, cost even more than they would in the developed world.

From the perspective of managing the sustainable development of these institutions, it makes economic sense for existing companies to pool resources and establish individual institutions able to perform some of these functions. Quite often, though, there is little impetus for doing so; competitive distrust in the developed world rarely brings investment consortia together, so why should it work in the developing world?

The times when consortia have worked in countries like the U.S. have been on occasions where the investment in research and development is too high, and when the public sector has been willing to take a role in assisting with the initial funding of the project. Expenses for billing and engineering support are analogous for a country like Russia, especially if institutions like the World Bank step forward and begin to help develop this needed service infrastructure for the telecommunications industry. There are presently a number of partnerships to fuel small business activity throughout Russia, most of which have been funded by foreign sources in cooperation with the Russian government. It will be possible for the Ministry of Communications to advocate that some funding and training be used to create small businesses that can undertake the marketing and sales functions, thereby decreasing the overall cost of service provision. In addition, the World Bank and other institutions can use their network of subcontractors and affiliated lending institutions to emphasize this opportunity and establish connections between the service providers and the emerging support sector for wireless access.

Co-Location of Facilities

The most straightforward means of establishing incentives to reduce the cost of services for consumers is to emphasize how the co-location of facilities has a positive impact on the balance sheet for wireless access providers. Clearly, the strongest argument for the co-location of facilities is that it keeps costs down, but in a highly competitive environment, many companies choose to forgo cost savings for the security of holding their own facilities. The result is the creation of a bottleneck, especially for wireless access; there are only so many appropriately positioned cell sites, and if they are occupied, it would be difficult for new entrants in the marketplace.

The bluntest public policy instrument is to mandate the co-location of facilities in specific circumstances, and look to translate successful models throughout the country as evidence of the business utility of co-location. As the Duma considers additional telecommunications legislation and the establishment of bodies for spectrum management, this kind of mandate would be an appropriate addition to the reform package. Given the difficulties in implementing policy, though, this is likely to be a difficult proposition to sell to corporate audiences unless appropriate incentives are built into the package. Although an in-depth discussion of taxation policies has not been a part of this analysis, there is much room for reform in this area, especially as it pertains to telecommunications providers.

CONCLUSIONS: POWER, AUTHORITY, AND SUSTAINABLE TELECOMMUNICATIONS DEVELOPMENT IN RUSSIA

The main theme of the discussion has been the need to establish a new pattern for authority in facilitating the sustainable development of the country's telecommunications infrastructure. The fragmentation of the telecommunications infrastructure, the lack of local investment, and the establishment and entrenchment of individual monopolies all seriously threaten the future viability of the telecommunications infrastructure. But none of that threatens the telecommunications sector in Russia as much as the lack of a strong, legitimate voice on telecommunications policy, strategy, and development that can affect change in corporate and political behaviors.

Nevertheless, the emphasis on the negative elements of Russia's recent development hides what could be turned into a very formidable strength. Most of the existing models of telecommunications development are "top-down," based on investment by large, state-owned or highly regulated, private monopoly service providers. The only "bottom-up" model we have been able to identify is the United States after the expiration of the Bell patent, before the creation of the Bell System and the Kingsbury commitment.

Russia has the potential of becoming the second major example of "bottom-up" telecommunications development. If the highly advanced pockets of telecommunications investment can be brought together through the coordination of influential groups within Russia (the banking community in particular), there is a real opportunity to establish a unified communications infrastructure that is more responsive to the long-term needs of customers. With local and regional investment driving the increased penetration, it will be possible to create a federation of telecommunications networks in Russia that achieve the goals of strategic liberalization that have been outlined.

In Russia, strategic liberalization can facilitate the integration of networks through establishing a common framework for investment and increased penetration. With different companies building wireless access networks and connecting them together to meet the increased demand, a consumer and mass market for telecommunications services will slowly develop. More importantly, the foundation for a competitive market will ground the future development of the sector. The result will be access competition, similar to the access competition that drove increased penetration in the U.S. from 1871 to 1913. So the fragmentation of the Russian infrastructure can be turned into an advantage, perhaps allowing it to leapfrog to a new stage of telecommunications development on a par with its competitors in the developed world. The troika driving Russia could not only avoid the ditch, but perhaps even overtake the other riders off in the distance.

ENDNOTES

[1] The research for this chapter was conducted in support of a Freedom Channel study on the development of Russia's telecommunications networks, sponsored by the World Bank and directed by Mikhail Kazachkov. Additional research was conducted in 1995 in cooperation with W. Russell Neuman, Franklin Miller and Shawn O'Donnell, and first appeared in "A Unified Communications Environment for Russia: White Paper on the Future of Russia's Telecommunications Sector," on file at the Edward R. Murrow Center at the Fletcher School of Law and Diplomacy.

[2] For students of the now defunct discipline of Sovietology, these facts put to rest once and for all the debate on the nature and structure of the Soviet system, with the case clearly showing the validity of the bureaucratic interest model over the totalitarian model of Karl Frederich and Zbigniew Brzezinski.

[3] One of the early examples of joint ventures in the equipment manufacturing sector was ItalTel, which was the investment vehicle for the Italian PTT. With the help of TsNIIS (Tsentral'nyi Nauchnoissledovatel'skii Institut Sviazi), under the leadership of Professor L. E. Verakin, ItalTel developed a new digital switch, the C-23. During this early period of foreign investment, both Alcatel and Siemens set up joint venture production facilities in the USSR. The Siemens venture announced in July, 1990 was a $95 million investment to manufacture EWSD and Soviet ESSDS switches with a Ukrainian company and the Soviet Ministry of Communications.

Establishing a New Procedural Entente: Strategic Liberalization in Brazil

From the perspective of national development, some of the problems that Brazil faces are common to many developing countries: disparity of wealth between population groups, a newly urbanized citizenry facing overcrowding, overpopulation, and poverty.[1] The size of the country, and the extensive (though often inaccessible) resources of Amazonia also set it apart, making Brazil a leader in the Americas by example alone. "As Brazil goes," President Richard Nixon once put it, "so goes Latin America."

But when it comes to telecommunications policy, Brazil has lagged behind in the region where it is thought to be a bellwether. While Chile, Argentina, Venezuela, and other countries have moved forward in the liberalization and deregulation of the telecommunications sector, Brazil has clung to the state-owned and operated company, Telebrás, that has run the country's telecommunications networks since 1972.

The overarching difficulty that Brazil faces in the telecommunications sector mirrors the stage of the country's development: Uneven pockets of success reside within an overbureaucratized, undercapitalized, and overburdened public telecommunications infrastructure. There is also the reality that, as is the case for many state-owned and operated PTOs, the political structure has guided the development of the sector to match certain institutional objectives that have little to do with rational economic decision making. Nothing short of a revolution in telecommunications policy could turn Brazil's present telecommunications infrastructure into a network to support national development.

On the surface, the comparison with Russia seems quite natural: Large countries with histories of authoritarian and totalitarian political systems, both have made the transition recently to a more democratic form of government. Events in both countries indicate that the political leadership has only a limited degree of political will to enact economic or social reform. Both countries are divided internally into regions with great degrees of authority and a disposition to act in accordance with local interests, not national ones. The trends seem to dictate that

223

a comprehensive program of liberalization and privatization, in both cases, are out of reach.

But, below that surface comparison, the differences are substantial. The country's birth as part of a colonial empire, then its incarnation as the center of that empire and the seat of the Portuguese monarchy, and finally as a modern Latin American nation state, set it apart from any of Russia's past historical experience. If any coherent strategy for telecommunications development is to be articulated, it will be expressed in the Brazilian fashion, summed up more than a decade and a half ago by Peter McDonough (1981), in his seminal work on the Brazilian elite structure, *Power and Ideology in Brazil*:

> The key question is not whether any one of the major ideological currents among the elites can dominate the others—whether, for example, the economists can gain sustained hegemony. Instead, it concerns the degree to which elites with profoundly discrepant priorities about the challenges facing Brazil, and themselves, manage to put up with one another. Such a compact is not identical to a consensus on specific issues, for the very concept of legitimacy involves the manner in which elites live with one another in the chronic absence of agreement about important policy questions. Neither, by postulate, is it reducible to the particular interests of disparate groups. It refers to a procedural *entente* rather than to a consensus on specific issues. (McDonough, 1981, p. 46)

As governments shifted from authoritarianism to democratic and back over the past century of Brazilian development, the need to define and implement a procedural entente has been critical. It is in defining the right "entente" that Brazil is able to move ahead in the developmental process.

The particular type of compact that stood at the foundation of Brazil's telecommunications policy over the past 20 years has failed to improve the telecommunications infrastructure of the country. In fact, as this discussion outlines in detail, that compact has undermined the viability of the network, leaving it even further behind the quality and capacity of the developed world than is the case in many neighboring countries.

As this book is being written, the Brazilian political leadership is struggling to find a new entente that can guide telecommunications development in the country for the next generation. Fernando Henrique Cardoso, with his right-of-center government, and his Minister of Communications, Sergio Motta, have been working to define an acceptable program to privatize the country's telecommunications networks by the end of 1998. The sale of portions of the wireline network have already begun, and more is planned in the months and years to come if countervailing political pressure can be overcome. Their efforts are constrained by the historical legacies of telecommunications development and the strength of the existing consensus, even in the face of increased competition and market advances through liberalization policies in all of the countries of Latin America.

This chapter presents our final case study, outlining some of the historical precedents that underlie Brazil's present telecommunications system and the emerging role of wireless communications as a driver of sector transformation. Areas of comparison with the other three case studies are presented, along with a discussion of what sustainable development through wireless communications may mean in the Brazilian context.

THE HISTORY OF BRAZILIAN TELECOMMUNICATIONS

The telegraph brought the first communications revolution to Brazil during the second half of the 19th century. E. Bradford Burns' (1993) description of the first push for telegraphy in Brazil says a great deal about the development of the telephone and other communications capabilities that were to follow.

The first line connected the imperial palace at São Cristóvão, on the outskirts of the capital, to the military headquarters at Campo Santa Ana in the capital. By 1857, Petropolis, the cool summer capital in the mountains behind Rio de Janeiro, communicated telegraphically with Rio de Janeiro. The outbreak of the Paraguayan war initiated a flurry of activity to string lines southward to the theatre of action. In a record time of six months a telegraph line connected the southern provinces to the court. (Burns, 1993, p. 154)

One of the clear realities in the history of development throughout Latin America has been the unmistakable influence of the military on the allocation of resources. Since the proclamation of the Brazilian Republic in 1889, the Brazilian Military has taken direct control of the major functions of government only once in 1964, but held it until 1985 (Keith and Hayes, 1976; O'Donnell, Schmitter, and Whitehead, 1986; Schneider, 1991).[2] Even so, the military has always been a critical interest group in the Brazilian content, claiming its share of resources and power as the nation developed.

The obvious importance of the military's influence in the telecommunications sector is underscored by the previous passage: The first telegraph connection was a military application; only then did it grow into a commercial service. That pattern is repeated numerous times in Brazilian history—a fact that has significant implications for the present attempts to revolutionize Brazil's telecommunications sector.

By 1855, the telegraph network covered 20,000 kilometers of lines, and in 1886, the first long-distance line connecting Rio de Janeiro to the southernmost state capital, Porto Alegre, was installed (again because of the needs of the military). The parallel of telecommunications development and military interest is especially pronounced in the 1960s and 1970s, defining, in part, the procedural entente that defines Brazil's present stage of telecommunications development.

Another part of the existing "telecommunications entente" in Brazil is the importance of international linkages over local ones. Burns' description of the development of the first international telegraph link also foreshadows some of our discussion:

> The progressive Viscounde de Mauã formed a company with an English partner to lay a submarine cable from Europe to Brazil. The idea of not only direct but instantaneous communication with Europe titillated the Brazilian imagination. A grand festival inaugurated that line. Seated before a special machine in the National Library on June 23, 1874, Pedro II dictated the first message to be cabled to Europe by telegraph long before it could similarly communicate with other parts of its own empire. The next stage in international communications was to establish telegraph contact with the Plata neighbours. The lines reached Montevideo in 1879 and Buenos Aires in 1883. (Burns, 1993, p. 155)

Brazil's historical link to Portugal was very much the focus of Brazil's relationship to the world toward the end of the 19th century. For a brief period of time before the collapse of the monarchy, Brazil was home to the Portuguese monarch after the Regent (and later King) John VI was forced to leave his home country during the period of French intervention in Spain and Portugal. Even after the Brazilian monarch split from the Portuguese dominion later in the century, the close connection between Brazil the colony and Europe the colonizers was reflected in the organization and penetration of the telegraph and telephone. That legacy still exists today, with international connections often easier to make than local ones. This is, in great part, because the present telecommunications structure rewards those kinds of investments more than investments in services in markets that could become broadly competitive.

Finally, there is Brazil's heritage as a country of some indigenous technological innovation and a market of interest to multinational companies. This factor was reflected in the first decades of the 20th century and the introduction of various telephone and radio systems throughout the country:

> The growth of the Brazilian economy in the first decades of the twentieth century and the expansion of government activities created a budding market for professional equipment, which was attended in the early thirties by at least a dozen national manufacturers, including Standard Electric S.A. (SESA) and the Companhia Marconi Brasileria. SESA was controlled by ITT, which also controlled the International Standard Electric Corporation (ISE), charged with the technical support to the automatic switching exchanges of the ITT European group in Brazil. Other major firms commercializing communications equipment in Brazil included Ericsson, established in 1891 to commercialize telephone exchange equipment and the Companhia Brasileria de Electriocidade, controlled by Siemens, which in 1913 installed several radiotelegraphic stations at Army forts (McKnight, Neuman et al., 1994).

Because of the potential size of the market and its distance from other major commercial centers, the local telecommunications equipment manufacturing base grew in the early decades of the 20th century through to the postwar era. This strength and tradition of telecommunications equipment manufacturing is reflected in the operating activities of Telebrás, Brazil's national telecommunications operator, during the last 20 years.

In many ways, these three factors formed the basis of the "telecommunications entente" of infrastructure development in Brazil. The influence of the military, the need for international communications and linkages, and the goal of establishing a local equipment and technology manufacturing base brought a variety of economic, social, and political interests together. The embodiment of that consensus is Telebrás, the state-owned and operated public telecommunications operator.

Early Developments and the Founding of Telebrás

The state did not take control of the telegraph or telecommunications infrastructure until late in its development. During the 1940s and 1950s, a number of multinational firms played a dominant role in the provision of wireless and wireline communications services, including telegraphy and telephony. The Companhia Telefónical Brasileria (CTB), which was controlled by a Canadian holding company, serviced around 70% of the 1.5 million telephones in the country in 1968 and carried about 80% of the telephone traffic. The CTB also was the main supplier of equipment for that network. Western Telegraph and Telephone, along with Cable & Wireless of Great Britain, dominated the market for telegraph services until the late 1940s, when the government began to increase its control over the sector through direct regulation. Concessions were given to four international carriers, all of which expired between 1970 and 1973, not to be renewed by the government.

At the local level, the initial pieces of what would become the Telebrás system had begun to emerge. In Goiás, for example, the first local telephone services were established in 1938 under the jurisdiction and ownership of the state government. According to the official history of Telegoiás, as it appeared on the company's website in late 1996, the state government had organized its holdings into the "Departamento Estandual de Comunicação," or DECO, by 1960. Over the next decade, the department underwent a number of changes, both in name and in stated purpose, leading up to the announcement of the Telebrás system's creation in 1972 and the final establishment of the system in 1974.

Unlike the American case, where lack of centralized control and private ownership of the telecommunications networks produced access competition, Brazil's somewhat chaotic map for telecommunications development did not lead to increased service penetration, at least when compared to the rest of the world. In 1957, Brazil's telephone density was a third of the world's average (1.3 per 100 inhabitants versus 3.7). In 1960, Brazil had 1 million telephones for a population of 70 million, with 2/3 of the equipment and traffic concentrated in the southern states of Rio de Janeiro and São Paulo, the industrial center of Brazil (McKnight, Neuman, et al. 1994). This underdevelopment of the telecommunications structure was a reflection of the pervasive economic difficulties of the country. After a booming decade of growth in the 1950s, increased dissatisfaction with the economic performance of the democratic regime, and the political confusion wrought by the quick succession of democratic leaders, the authority of the government disintegrated (Burns, 1993). The military coup of 1964 ushered in an era dominated by central (and military) prerogatives, a fact reflected in the next era of Brazilian telecommunications development.

As the 1960s progressed, the Brazilian military government began to assert authority over the various pieces of the Brazilian economy and society, instituting a system of authoritarian control that would last through the 1970s. The assertion of central control included the reorganization of the country's information and telecommunications networks and the creation of a single, government-owned telecommunications operator: Telebrás.

The first step to establishing a central authority for telecommunications operations and development was made in 1962, with the passage of Brazil's telecommunications code. "The Code granted the state a monopoly in the operation and regulation of telecommunications activities," and established a National Telecommunications Council (CONTEL) to reduce market fragmentation and rationalize equipment supplies (McKnight, Neuman et al. 1994).

Soon after the completion of the military coup, the Brazilian Telecommunications Enterprise (Embratel) was created. The goal of Embratel was to "integrate the vast territory of Brazil through telecommunications, linking the country to the rest of the world" (Embratel, 1996). The company's mission, even to the present day, sounds almost messianic. As Embratel's official history as it read on the company's website at the end of 1996 put it:

> The first historical example of a worldwide telephone link took place just a few days after the inauguration of the Tangua earth station in Rio de Janeiro State, transmitting the blessings of Pope Paul VI directly from Rome by satellite in 1969. This same year, Man first set foot on the Moon, a benchmark that was eclipsed only by Brazil's third World Soccer Championship, broadcast in full color the following year. Then came Direct Long-Distance dialing, facilitating the lives of millions of people in need of better telecommunications and interlinking countless towns and cities all over Brazil.

The initial objectives of the company closely mirrored the military's desire to assert central control over the fragmented political and economic system.[3] Embratel took charge of the national and international trunk operations and the development of linkages with the farthest reaches of the country. A Ministry of Communications was established in 1967 as part of a broader government reorganization that marked the military's entrenchment in the government and the opening of a period of political cohesion and terror. As part of a Ministry of Communications rationalization effort, the Brazilian government acquired the largest foreign telephone concessionaire, the CTB, and began to take direct control of most of the telephone companies in the country.

In 1972, the government announced the establishment of Telebrás as a public enterprise "to plan and manage, financially and technically, the development of the [telecommunications] system." Through a series of purchases and mergers, Telebrás gained majority control of the country's telephone and telecommunications networks, reducing the total number of major networks to 37. The government owned 80% of the enterprise, with Bradesci, Brazil's largest private bank, AT&T, and Bell Canada as minority private shareholders (Sussman and Lent, 1991). Subsidiary firms were established to serve the commercial and government interests of a given state, and the federal structure of Brazil was such that state authorities were able to maintain the management control of the state enterprises. But because Telebrás was given a monopoly on the purchase of telecommunications equipment, it had the ability to dictate the technology policy and infrastructure development of the state companies.

As part of this power to purchase equipment, and in executing its mandate to develop a national telecommunications research and development strategy, Telebrás founded the "Centro de Pesquisas e Desenvolvimento"—the Center for Research and Development, CPqD—in 1976. CPqD was responsible for the development of technology and the training of Telebrás personnel so as to ensure the quality of the telecommunications network. As such, it became the national center for technology research and development, and Telebrás R&D activities through CPqD have accounted for over 90% of all Brazilian telecommunications research. It also became an example for the world during the 1970s and 1980s as countries looked to develop centralized strategies for national development.

With the consolidation of the telecommunications sector complete and the structure set for the development of the sector, Telebrás and the Brazilian government set about implementing a centralized investment and growth strategy. The centerpiece was the Second National Development Plan, with a goal of a 200% increase in the number of telephones in use by the end of the 1970s. Spending on telecommunications investments went from just over $500 million in 1973 to more than $1.5 billion in 1976, all of which were funded by a central government flush with cash and new international loans.

The strategy for research and development followed the pattern of infrastructure investment in telecommunications technology. As Michael Hobday described:

> One can identify a three-pronged approach to building up the technological and industrial base of the country: First, the setting up of a major new government-owned center in digital technology; second, to 'persuade' the MNCs to transfer the ownership to Brazilian capital and to increase the level of manufacturing and technological facilities within the country; third, to sponsor the development of new Brazilian companies to manufacture equipment and develop technology. (Hobday, 1988, p. 88)

The classic strategy of import substitution was put into place, and CpqD became the fulcrum for partnerships between multinational and local firms to supply the infrastructure for telecommunications development. The Ministry of Communications saved percentages of the market for different kinds of technology for local companies, while the remaining percentage could be manufactured locally by the multinationals (McKnight, Neuman et al., 1994). Even so, the multinationals were compelled to lease from CpqD the technological specifications so that they could produce the equipment, thereby further fueling the allocation of resources for the research center.

The result was a re-establishment of foreign presence in Brazil according to the new parameters:

The rapid growth in public telecommunications investment and Telebrás standardization policies prompted foreign firms active in the Brazilian market to set up Brazilian manufacturing operations such as Siemens (1970), Phillips (1974), Ericsson (1955), SESA (ITT) (1967), and NEC (1968). In the period between 1971 and 1974, investments in telecommunications infrastructure reached $588 million and Telebrás orders for about 1 million lines were divided between Ericsson, SESA (ITT), and Plessey. (McKnight, Neuman et al., 1994)

Each of those companies established an association with Brazilian financial groups in order to become a "nationalized" part of the telecommunications sector (Capallaro, 1993).[4] The sector as a whole grew through this process of centralization and rationalization, with the equipment manufacturing capacity of Brazil's economy increasing to 1.3 billion by the end of the 1980s. Most of this technology supported advanced digital transmission, and superseded the old analog transmission systems of the previous era.

Hobday described this as the real victory of the centralized development model and import substitution strategy. "If we contrast the U.K. with Brazil" he wrote, "the ease of transition to digital technology and the leapfrogging of various intermediate forms of telecommunications technology is striking." Hobday pointed to the development and deployment of the digital Tropico switching system through the Telebrás research system as a direct transition from an analog to digital structure. At the same time, the U.K. had to pass through a variety of smaller steps to begin the process of transition during the 1960s. The result, when looking from his vantage point in the mid-1980s, was a Brazilian infrastructure with a higher penetration of digital technology than most other developing countries—even though the absolute penetration of services was lower.

Nevertheless, these successes masked a broader failure in the telecommunications development strategy, one that became more obvious as the 1980s progressed. The failure occurred because this procedural entente of centralizing power and rationalizing systems did not produce the sustained growth of the telecommunications system. Quite the contrary, it set the preconditions for the erosion of the financial and technological integrity of the Telebrás system.

The Failure of the First Brazilian Procedural Entente

At its core, the problem with the centralized system for the development of the telecommunications networks was that it depended almost exclusively on the vicissitudes of national economic prosperity. A succession of economic problems, including the two world oil crises and bouts of serious inflation, undermined the macroeconomic stability of the Brazilian economy. The impact on state-run firms in general was significant:

> State enterprise investment outlays had already begun to drop in 1982, a year before the generalized decline in Brazil's investment ratio became evident. This decrease, from 6.1 percent of GDP in 1981 to 3.8 percent in 1984, reflected in a variety of factors, including the following: The acute shortage of foreign exchange; constraints on foreign and domestic borrowing; falling revenues because of price controls and because of weak demand for selected goods and services produced by the public sector (for example, steel and electric energy); and numerous import restrictions and prohibitions. The decline in private sector outlays was also steep, as the investment ratio (excluding the state enterprises) dropped from a 14.6 percent average of GDP in 1981 to 10.8 percent in 1984. (Dinsmoor, 1990, p. 5)

The decrease in consumption caused an economic contraction, one that affected the telecommunications infrastructure very seriously. The nature of the impact was particular to the telecommunications sector, and the problems it engendered are, in many ways, peculiar to the Brazilian case. To put it simply, the attempt by the center to hold onto the procedural entente that guided the development of the telecommunications networks through the period after 1964 increased the damage done to the sector.

The next few pages offer statistical information which shows, in detail, the impact of the Brazilian economic contraction on the telecommunications sector. The numbers clearly show investment in the telecommunications infrastructure was not sustained because of structural and institutional deficiencies in the public sector. .

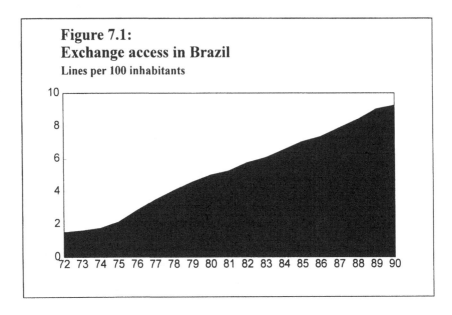

Figure 7.1:
Exchange access in Brazil
Lines per 100 inhabitants

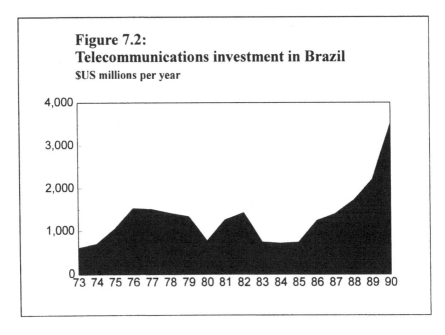

Figure 7.2:
Telecommunications investment in Brazil
$US millions per year

Note: Figures 7.1 and 7.2 are adapted from company data and International Telecommunications Union (1994, 1995, 1996)

Figure 7.1 shows that the number of telephone exchange access lines increased at a consistent rate during the period between 1972 and 1990. From a total of about 1.4 million, the number increased to a little more than 8 million lines. The increase was accomplished by government mandate, with targets for expansion set mostly by the political agenda of further expansion and centralization. On the surface, these numbers seem to indicate a successful strategy for the development of the sector, with increasing capability and access for the citizens of the country.

But even as the number of lines increased at a steady pace, the money available to construct and support those lines was decreasing during the 1970s and into the 1980s. Figure 7.2 shows the level of investment in the Brazilian telecommunications infrastructure during the same period of 1973 to 1990. The vertical bars show the total investment in millions of dollars

As the Telebrás system consolidated in the mid-to-late 1970s, there was an increase in investment, but the total investment decreased dramatically from $1.7 billion in 1976 down to about $1 billion in 1980. Investment did not reach the 1976 level again until more than a decade later, as the Brazilian economy made the transition back from dictatorship to democracy. In other words, there were more lines to operate and maintain with less money, which was a recipe for a deterioration of the system.

Considering the decrease in the gross consumption of the government throughout the 1980s, it might be assumed that all of the public and private enterprises shrunk at approximately the same rate. Chart 7.3 clearly indicates that the telecommunications sector, relative to other sectors as a total share of GDP, decreased dramatically during this period of contraction. It shows the level of telecommunications investment in Brazil as a share of GDP between the years of 1976 and 1992. In 1976, at the high point of telecommunications investment in absolute terms, the sector was responsible for a little more than 1% of GNP. That number drops to about .5% of GDP through the 1980s. Other sectors of the Brazilian economy were able to gain access to a greater percentage of resources relative to the telecommunications sector, even while the level of telecommunications investment measured as a percentage of GNP in most other countries increased dramatically during the 1980s.

Even as the economic downturn forced a contraction of resources for Telebrás, the demand for telecommunications services continued unabated. Instead of turning away customers, Telebrás made sales and instituted a system whereby subscribers would pay for the construction of the access line. The system put into place asked subscribers to make a contribution of equity capital in advance of receiving services, but, as a result of the high demand, many made the investment and waited for months or years to receive their telephone line (Saunders et. al., 1993).

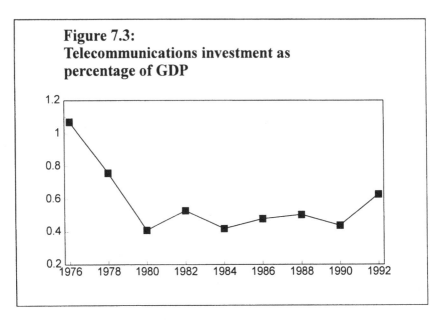

Figure 7.3:
Telecommunications investment as
percentage of GDP

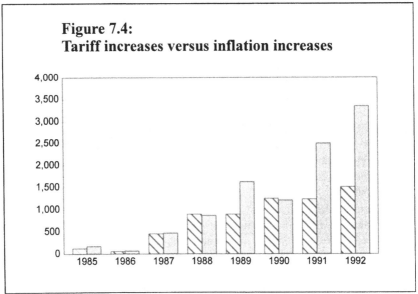

Figure 7.4:
Tariff increases versus inflation increases

Note: Adapted from company data and McKnight, Neuman et al, 1994.

Actual procurement of lines lagged behind sales for most of the period between 1975 and 1984—in 1980, for instance, more than 600,000 lines were sold, but less than 400,000 were actually constructed. The result of this policy of selling lines before they were built was the center could not handle the demand. Huge numbers of customers were not being serviced by the lines they had already "purchased," and the ability of the network to maintain the same quality of service was seriously hampered.

With prices spinning out of control, the political pressure was great to maintain a low cost of service for subscribers throughout the country. Figure 7.4 shows that Brazil's government fell prey to what is a common story in telecommunications development: Increases in prices lagged well below increases in inflation, thereby eroding the real return on telecommunications investment. This figure compares the annual tariff increase allowed by the government since 1985 with the annual inflation rate. In 5 of the 8 years, the tariff increase was substantially lower than the increase in annual inflation. In total, during the 1980s, telecommunications tariffs declined by 80% percent when adjusted to ever-rising inflation.

It comes down to basic business economics. If the telecommunications operator's revenues cannot keep up with inflation, there is a tremendous disincentive to invest, thereby undermining the capabilities and maintenance of the telecommunications network. By 1990, there was an estimated $2.4 billion dollar gap in the Telebrás budget, and the funds necessary to deliver the 1.2 million terminals already sold but not yet delivered was unavailable. "In less than a decade," McKnight, Neuman et al. (1994) write in their report, "the government subtracted about $7 billion dollars from Telebrás' investment capability."

But the way in which Brazil brought itself into this difficult situation is closely tied to the progression of authoritarian politics and the transition to the government of Jose Sarney in the late 1980s:

> After the institutional consolidation of the seventies, the Telebrás system fell prey to the politics of the democratic transition of the early eighties, which increased the power of state governments and the Congress. Little by little, the professional management and technical staff of Telebrás, already threatened by short term financial management, became victims of widespread clientelism, with high levels of inefficiency and mounting bureaucratization (traditional public service agencies). This pressure on Telebrás intensified after 1985, when the Sarney government replaced top level professional management by political appointees, part of a larger game of political manoeuvring. The stagnation and decline of the Telebrás system in the eighties was part of a larger deterioration of the existing governmental regime and policy framework, and the attendant difficulties of fashioning a functional democratic system, with effective checks and balances, out of the ashes of the discredited military/authoritarian system. (McKnight, Neuman et al., 1994)

The system that protected Telebrás through import substitution and centralization was now undermining its ability to act and function as a provider of telecommunications services. Part of the problem was that the centralization of telecommunications providers reduced the opportunities for independent and private investment. There were no alternate institutions in place that could pick up where Telebrás left off and continue to develop the networks and infrastructure.

The initial deployment of fiber and digital technology had put Brazil ahead of other developing countries, but the penetration of those kinds of capabilities are limited. Michael Hobday wrote:

> According to the Brazilian census of 1980, the total number of terminals in the rural sector amounted to only 75,000. In 1983, this figure stood at little more than 90,000, or, put another way, only 1% of rural households possessed a telephone. This extremely poor level of telephone coverage reflects, on one hand, the investment restrictions imposed by the government and, on the other hand, the much deeper poverty of the rural areas in relation to the main industrial centers. (Hobday, 1988, p. 106)

There were still problems at the periphery, even after the relative substantial investment of the early to mid 1970s. The erosion of financial resources during the 1980s did not help matters much, and hampered Telebrás' ability to meet a critical long-term objective: Providing telecommunications services to the country.

The Present Structure of Brazil's Telecommunications Sector

The structural basis for Brazil's telecommunications entente from the past two decade appears in the next chart. Telebrás has been put in place as the coordinative and structural unity that the military government required in the 1960s and 1970s, but the decentralization of the system has left a number of gaps in the system to satisfy the local interests of various governments and groups.The core of local autonomy is housed in the pole companies, which are organized within each of Brazil's state.

As Figure 7.5 indicates, the top five pole companies account for more than 53% of all revenues. The largest, TELESP, is responsible for more than a quarter of Telebrás revenues. Embratel tends to be less autonomous than the individual pole companies, but that is not because of a less significant role in producing Telebrás revenues; Embratel is presently responsible for approximately 24% of all Telebrás revenues.

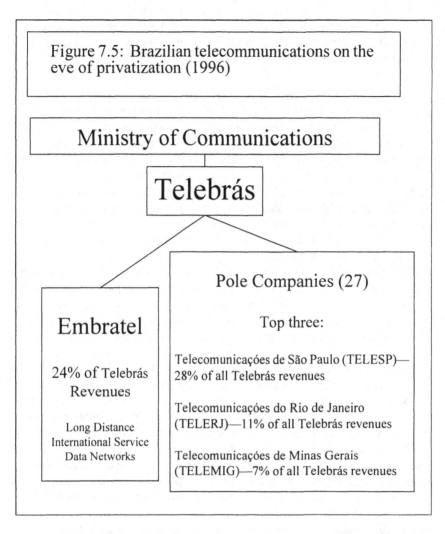

Figure 7.5: Brazilian telecommunications on the eve of privatization (1996)

Ministry of Communications

Telebrás

Embratel

24% of Telebrás Revenues

Long Distance
International Service
Data Networks

Pole Companies (27)

Top three:

Telecomunicaçóes de São Paulo (TELESP)—28% of all Telebrás revenues

Telecomunicaçóes do Rio de Janeiro (TELERJ)—11% of all Telebrás revenues

Telecomunicaçóes de Minas Gerais (TELEMIG)—7% of all Telebrás revenues

The tension between those institutions can be seen in the constitutional wrangling that has taken place since the transition to democratic government in the late 1980s:

> The 1988 constitution entrenched the above structure in the new democratic law. The 1988 Constitution, in its chapter entitled "On the Organization of the State," makes explicit the exclusive competence of the Union. In regard to public telecommunications services, paragraph 11 states that "the Union directly or through authorization or concession to an enterprise under state share holding control, the telephone, the telegraphic, and data transmission services and other public telecommunications services, insured the provision of information services by private law entities through

the public network of communication." In paragraph 12, [the constitution] states that companies can "explore directly or though authorization, concession or permission, broadcasting services, sound and image services and other telecommunications services." (McKnight, Neuman et al., 1994)

One reading of the preceding extract would indicate that a monopoly system is in place and that the public network controls the provision of all telecommunications services. On the other hand, there is enough room to formally recognize competition, but it is clear that a great deal of competition would not be allowed. In one regard, the passages reflect the tension between the centralized control of Telebrás and the regional autonomy of the pole companies who, in many cases, have begun to go their own way in providing services and building infrastructure to meet demand. Telepar, the Paraná state telecommunications operating company, has been investing its own funds to set up new transmission lines and promote data communications services and local communications networks—a market which traditionally has been left to Embratel.

The Collor administration, which took office amid much fanfare in 1990 and left in disgrace before the end of its term, spent a great deal of its legislative energy on proposing alterations to the 1988 Constitution. This is also true with regard to the telecommunications market. As the government began to roll out its privatization program, congressional interests were able to point to the Constitution to protect sensitive public sector companies, Telebrás included. The Collor administration did open the telecommunications market to limited competition in paging, cellular, cable television, infrastructure provision, and private data networks as part of the presidential decree entitled "The Regulation of Limited Services" in July, 1991. But it was impossible to go further in liberalization of the telecommunications because of the proscription of the Constitution.

The Collor administration quickly fell into a cycle of corruption allegations and political gridlock. After Collor's resignation in the midst of charges of serious corruption, interim president Itamar Franco took up the mantle of constitutional reform. More than 16,000 separate reform proposals were made, many of which would have directly or indirectly affected the government's role in the telecommunications industry. But very little was done; of the 113 congressional sessions scheduled to tackle the reform proposals, only 71 took place, in great part because of persistent boycotts and protest by congressional leaders. Needless to say, this period strained the already difficult relationship between the government and senior executives at Telebrás, who were facing consistent political pressure to increase infrastructure investment even when the political process was unable to offer the company the money required to complete those investments. In 1994, Telebrás president Adyr Silva publicly complained that the company would have to spend $2.5 billion "just to fulfil President Franco's requirement to expand private and public telephone lines. This amount alone almost absorbs the total of 2.9 billion dollars allocated for Telebrás this year [1994]."

The hard fought presidential campaign that took place against the backdrop of this caretaker presidency pitted a leader with a long tradition of labor support, Luis Ignacio da Silva, against a former government finance minister, Fernando Henrique Cardoso. Cardoso began the campaign far behind da Silva but was buoyed by his successful anti-inflation campaign, which played a critical role in the re-stabilization of the Brazilian economy in 1994. His victory in the first round was considered by many commentators an opportunity for the further liberalization of the telecommunications market; his support for the stillborn privatization program of his predecessor would definitely transform the telecommunications sector.

In June, 1995, the Brazilian government began to take the difficult steps toward privatization and liberalization. The lower house of the Brazilian Congress voted 357 to 136 to approve a constitutional revision to open up the telecommunications market. The provision removes the Telebrás monopoly in the marketplace and gives the government the latitude to privatize the company, should it choose to do so.

But from that point on, the road to privatization has been bumpy. Initial provisions for changing the legal basis for the government monopoly have been challenged in the upper house of the Congress, and the government appears more interested in spending its political capital on privatizations in other sectors before turning to implement the more difficult proposition of full privatization and liberalization for Telebrás.

Almost a full year after the government's first success in the Congress, the Ministry of Communications revealed its tentative plan for the full privatization and liberalization of Brazil's telecommunications market. According to information available on the Telebrás website soon after the announcement, the following elements formed the basis of the proposal:

- Finalization of constitutional reform, permitting private investment in fixed telephony, cellular, satellite, and value-added networks;

- Licensing of the "B" band cellular services, which has been delayed and has undergone various difficulties and challenges from private sector companies wishing to be involved in this portion of the market;

- Privatization of band "A" cellular holdings, presently owned and operated by the pole companies as part of the Telebrás system;

- Opening the market for satellite services;

- Creating a new General Law for telecommunications services to replace the existing law, 4.117, which was put into place in 1962;

- Establishing a regulatory body to oversee the privatization process, as well as facilitate arrangements on issues like interconnection and ensuring competition among state operators and private concessionaries; and,

- Dividing Telebrás into five or six "mini-holding" companies by geography in advance of privatization.

The present goal is to complete privatization by the end of 1998, and the sale of various components of the Telebrás system began officially in late 1996. Even so, the above list seems ambitious at best, but parts of the program are achievable in the near term—if the legal and political barriers can be removed. But that will be a very difficult task, requiring the transformation of the existing telecommunications entente.

The Existing Telecommunications Entente of Brazil

The initial tensions surrounding the privatization of Telebrás and the liberalization of the market show both the strengths and the weaknesses of the telecommunications entente that has defined the development of Brazilian telecommunications sector since the military coup of 1964. Like the authoritarian system imposed by the Brazilian military, the Telebrás structure imposed on the telecommunications networks sustained itself for a period of time by limiting the tension between local and national policymaking. The telecommunications entente of Brazil was that the center would provide the advantages of research, development, and investment, and the local companies would be able to maintain a reasonable level of autonomy.

But technology and global economic change has undermined the entente, a fact clear in the investment patterns of the Brazilian government in the past decade. Such a centralized, top-down model for investment and telecommunications management cannot provide for the telecommunications needs of the country. The lack of performance, in that regard, undermined the entente and now forces the local, national, and various economic interests to carve out new relationships.

The details of the liberalization and privatization program change as this book is being written, but it is certain that even if Telebrás were to be privatized today and the most liberal telecommunications regime in the world were put in place the next day, the significant patterns of corporate behavior would not immediately change. If Brazilian politics continue to form, what will develop is less a set of rigorously applied rules than a set of norms and dynamics that spell out another procedural entente to satisfy the perceived needs of development for the next generation of social change.

The goal is not to predict the nature of the entente that will develop from this present time of change, but rather suggest the ways in which strategic liberalization, as a policy that links corporate and government interests, can contribute to the achievement of the ultimate goal of telecommunications development: increased service penetration and improved quality of service. By introducing a new dynamic of extensive competition and investment in wireless access services, Brazil can focus resources on a appropriate technology that can become the backbone for future sector development.

The next section discusses the history of wireless access technology in Brazil, focusing on some of the country's efforts to link the huge expanse of the country through microwave systems, satellites, radio, and other forms of wireless communications—including the recent introduction of broadly available cellular services. The final section deals with to some points of comparison and the factors that are likely to determine the success and failure of the government's efforts to transform the country's telecommunications sector.

WIRELESS COMMUNICATIONS IN BRAZIL

The history of wireless communications in Brazil is marked by the adoption of existing models of wireless and a lack of innovation on the part of wireless access providers. In the realm of satellite and cellular, Brazil has chosen to follow the well-worn path of the developed world, instead of focusing on how cellular access could be used to push levels of penetration higher throughout all the regions and economic strata of the country.

Wireless access began, as it did in most developing countries, with the introduction of radio communications. The first shortwave transmitters were installed in Brazil during the late 1920s by Radiobras, a wholly owned subsidiary of RCA International of Paris. Fifteen radio stations were built during this initial period of growth, creating incentives for the emergence of a local radio industry.

Although the Brazilian radio and broadcast television services would grow to be one of the largest in the developing world, other forms of radio access were late in being introduced throughout Brazil. The first use of wireless access involved satellite and microwave applications in the 1960s and 1970s, with ESMR and cellular following in the 1980s.

Satellite Telecommunications:
Information Solutions for a Large Country

Like many countries attempting to overcome the problems of geographic dispersion and diversity, Brazil turned to satellite to bridge gaps which would have been uneconomical to link through traditional wireline architectures. But satellite technology has only had a limited impact on the telecommunications development of the country. Satellite has done more to enhance access for transnational communications than to increase penetration of access services. In that regard, satellite has yet to contribute in a broadly positive fashion to the telecommunications development of the country.

In Brazil and throughout the world, INTELSAT was the main initial influence shaping the introduction of satellite technology. In fact, Brazil was the first country to become a signatory of the treaty that brought INTELSAT into existence. Embratel did not begin leasing INTELSAT transponders for long-distance use until 1978, choosing initially to employ point-to-point microwave in an attempt to reach the outer edges of the country. But the satellites were needed to reach into the vast Amazonia region, which became more and more the focus of resource extraction and economic opportunity during the waning days of the military government.

Embratel soon decided on a bolder strategy, the construction and launch of its own satellite network. The $210 million dollar program culminated in the launching of Brazilsat I and II in 1985 and 1986 respectively. Embratel rented the satellite's transponders to various intensive users of telecommunications services, and the pattern of usage was what would be expected: "The main beneficiaries of satellite technology were commercial television and transnational corporations" (Sussman and Lent, 1991).

That was not because of a lack of proposed social uses for satellite technologies. For example, a transponder from Brazil's domestic satellite service was proposed as the national transmission backbone for the Ministry of Education's FUNTEVE network in 1986, but the use of the satellite "would have increased the network's transmission costs by a factor of ten" (Saunders, Warford and Wellenius, 1993). Even though two more satellites were planned for launch in the late 1980s, the civilian government reduced the funding for the telecommunications infrastructure, and the Brazilian satellite development program was one of the first to be cut back. It has failed in its broader social purpose: improving access of information resources to all of the people within the country.

The pattern of international usage and highly priced applications has continued through a pattern of smaller and medium sized investments in satellite access. Telebrás, through its Embratel subsidiary, has continued to link up to a number of proposed international telecommunications ventures that employ satellite technologies. They include a partnership with two separate consortia and another with Bell Atlantic.

From the perspective of creating an indigenous market for satellite equipment and services, Brazil did not achieve any lasting achievements for the development of the telecommunications sector. Brazilsat relied heavily on contributions of finance and technology from American satellite firms. Even with the strength and resources of a CPqD, it took Brazil "about 20 years to implement indigenized adoption of [satellite technology] for domestic uses," well behind the quick adoption rates of other developing countries (Sussman and Lent, 1991). The equipment brought into the country to support VSAT networks and data communications is predominantly from countries in the developed world. Generally, they are focusing on developing the market among local industrial groups with the resources and interest to fund this kind of telecommunications investment.

Cellular and SMR: The Development of the Commercial Sector

The development of satellite access in Brazil followed the established pattern seen in many developing countries. The same can be said of cellular and Specialized Mobile Radio (SMR). Both kinds of services developed in response to the models already established in the United States and other areas of the developed world, and have been constructed more to serve the needs of the well-to-do few than the broader market demand for telecommunications services.

Because of the great distances for travel and transportation throughout the country, one of the first wireless applications for mobile telephony was a emergency service that worked in much the same fashion as a fleet dispatch service. Transport companies were provided a service through the Telebrás system that allowed their drivers to access the national phone network from anywhere on the road, using a very high frequency, manually switched radio telephone system. This early version of Specialized Mobile Radio (SMR) has become commercialized in recent years, with companies such as Comcast taking a position in the development of this market after licenses for SMR services were granted in 1993 and 1994. But it has not been turned into a broadly available wireless access service, in great part because SMR as a whole has not been able to establish itself as a direct competitor to cellular in the developed world, much less a lower cost wireless access system with higher penetration potential.

Embratel, in turn, has focused some of its attentions on the fleet dispatch market, using its satellite technology to compete in this growing market. The company's Movsat C service is positioned as a guaranteed communications link between "vessels, vehicles, offices, work-yards etc. using small antenna and low-cost receivers" (Embratel, 1996).

But the most notable story of wireless access in Brazil over the past 2 years has been the story of cellular telephony. The story lies not so much in the growth of cellular telephony as in the political wrangling behind the definition of the markets and the eventual structure of licensing. Considering the global boom in cellular services and wireless access, the debate over key questions took on special significance: Who would control the licensing procedure? Would licenses cover national or regional areas? Would Telebrás or the pole companies have the direct responsibility over the operation of the licenses?

The final result was the following: The market is divided into two bands, following the duopoly pattern established in countries like the U.S. in the initial phases of cellular development. Band A has been allocated to the local pole operating companies, and Band B will be assigned to the private sector. In the private service competition, Telebrás makes a technical pre-selection, and the local pole company makes the final decision based on financial criteria. Foreign firm participation is limited to 49% of the shares, although supply and installation of the system can be contracted out. To this end, joint ventures have been established between foreign technology providers and Brazilian service providers, including some of Brazil's largest industrial and financial groups.

The pole companies have clearly been in the driver's seat when it comes to the development of Brazil's first cellular systems—in fact, investment patterns among the pole companies seem to indicate that cellular services quickly have become much more important than the fixed networks being managed by the company. Telesp, the biggest of the pole companies, put in service 470,000 lines in

1995. According to the company, 256,000 of those lines were fixed link and 214,000 were cellular (Telesp, 1996). The pattern is replicated at Telerj. Of the U.S.$885 million invested by the company in 1995, more than 44% went to building the company's cellular networks, a greater percentage than went to supporting access through conventional, fixed link services (Telerj, 1996).

The impact on revenue streams is evident; Telepar claims more than 60,000 subscribers on its cellular system as of March, 1996, with Figure 7.7 depicting the breakdown of company revenue from its total of more than 207 RS Million. Ten percent of the company's total revenue is already coming from the company's cellular system, and this is after only a year of operation. Between the increased levels of investment, and the apparent impact on the balance sheets of the pole companies, there is clearly an explosion going on in the country's cellular telephone market.

But what kind of explosion is it? A quick history of pricing strategies since the inception of cellular services makes the situation a bit clearer from the perspective of telecommunications development. In 1993, the first cellular system came on line in the city of São Paulo. Access fees for initial hook-up were $2000, with the air time costing an average of $.40 a minute and the terminal costs between $550 and $3,000. Clearly, the pricing reflected a traditional cellular strategy of offering high mobility at a high cost to consumers. Considering that the per capita income of the country is only $2,770 per year, it is hard to imagine that many Brazilian citizens would be able to afford these new services.

Even with the steep prices, Telebrás estimated that there would be 600,000 cellular phones in operation by 1994 and a potential market of 1 million terminals by 1995. An investment requirement of $1.5 billion was estimated, $500 million of which would be for network equipment. A more recent estimate puts the total number of subscribers at 721,000, while an estimated 1.5 million people are on waiting lists for cellular phones.

The present tariff structure put in place by the various pole companies looks suspiciously like those in countries like the U.S. or the U.K.. For example, Telerj's service connection costs R$40 or R$47, depending on the kind of service desired. At the time of this writing, a real is about the same as a U.S. dollar, so the numbers should sound very familiar to anyone who owns a cellular phone in the U.S.. Depending on the service plan, the per-minute rate for a call in peak hours ranges from R$.40 to R$.91, and per-minute charges for nonpeak hours range from R$.28 to R$.76. Again, the numbers sound very similar to those charged in the U.S.— and this is a country with a significantly lower per capita income rate. The explosion, at least in terms of the pricing, is the explosion of the cellular model, again being exported to developing world countries that cannot support sustainable growth in the telecommunications sector through such a strategy.

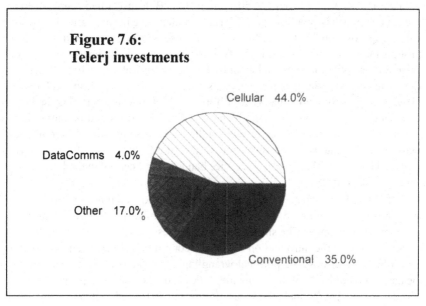

Figure 7.6:
Telerj investments

Cellular   44.0%

DataComms   4.0%

Other   17.0%

Conventional   35.0%

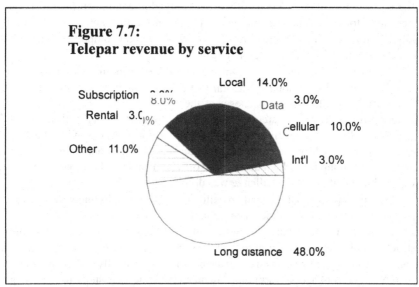

Figure 7.7:
Telepar revenue by service

Local   14.0%

Subscription   8.0%

Data   3.0%

Rental   3.0%

Cellular   10.0%

Other   11.0%

Int'l   3.0%

Long distance   48.0%

Note: Figures 7.6 and 7.7 are adapted from company data available on Telerj and Telepar websites in September, 1996

Perhaps the pricing would be different if Band B licensees had been able to aggressively launch competing services. But, to date, those licenses are still awaiting official approval so that the bidding for the licenses can begin—and, if the bidding process takes as long as it did in the U.S. for PCS, it is likely that the Band B cellular systems will not come on line until the end of 1997 at the earliest. When they do come on line, one group of analysts from Donaldson, Lufkin, and Jenrette (1996) estimate that more than U.S.$1 billion will be spent on acquiring the licenses even before network construction begins. These realities and potential delays in the cellular licensing procedure forestalls even opening a discussion about licensing higher bands of the spectrum, such as the PCS bands between 1.8 GHz and 2.2 GHz or the fixed wireless service bands being licensed in the U.K., to provide further competition.

And this halting pattern of telecommunications development is further complicated by the politics of privatization and the historical involvement of the equipment companies. The shared selection process, along with the need for partnerships, slowed the introduction of cellular communications systems by opening a whole range of possible political wrangling. The technical preselection was to be based on the ability of the companies bidding on the license to provide a system similar to the AMPS analog standard in use in the United States (Tarlin, 1994). The partnerships of choice generally brought together equipment producers with local investors. Considering the deep tradition of conflict between the various equipment producers, and the close links between certain companies and the political system, conflicts were bound to arise.

One of the more interesting conflicts pitted Ericsson and Motorola against NEC do Brazil. Even though Ericsson had a long established relationship with Telesp of São Paulo, NEC was chosen as the partner for the company in the construction of the cellular system. Ericsson and Motorola immediately took NEC and TELESP to court, claiming that NEC could not in fact provide the IS41B equipment required for the construction of the system. The companies eventually won a court injunction, which delayed the official licensing of the system by some weeks, but eventually the challenge was dropped.

The initial phase of the Brazilian cellular system's development has patterned itself closely on the U.S. model. The technology is AMPS, the pricing is a bit higher than in the U.S. but patterned in a similar fashion, and the initial duopoly is structured like cellular's beginnings in the U.S., with the regional telephone company receiving one license and a private company (eventually) to receive the second. Yet wireless access forms the foundation of the first steps in the recently announced privatization plans. What is the source of this seeming contradiction? The telecommunications entente in Brazil still owns wireless access, both as a business proposition and as a regulatory definition. Breaking free of the system, and creating the conditions whereby wireless access can evolve into a driving force in the country's telecommunications development, will be a difficult task.

Breaking Free of the System:
Evolutionary Paths for Wireless Access in Brazil

Considering the pricing strategies, the political conflicts, and the still open question of the privatization of Telebrás, it appears that the establishment of a strategy to push the penetration of wireless services is a long way away. The extended battle for the licensing of cellular systems and the potential for open conflict between the pole companies and Telebrás over control of certain strategic assets brings to the center the critical question of the Brazilian case: What institution can carry strategic liberalization forward?

In the U.S. and U.K. cases, liberalization through wireless communications appears to be possible only through action by government regulators in backing away from the stringent definitions and restrictions placed on wireless access providers. Establishing a pathway for the convergence of wireless and wireline access will depend on policymakers carrying the liberalization path forward to allow the multiple providers now entering the marketplace sufficient freedom to grow, fail and be reborn through the evolutionary process. The evolutionary process would be one of transformation from niche services to that of a broad access technology—the growing pains would be substantial, but it is quite likely that service providers would evolve and thrive in such an environment.

In Russia, the free market exists almost by default, and is vigorously exercised at the local level. Performance in telecommunications development through wireless access is, therefore, widely varied according to the quality of local political and economic leadership. The concern in the Russian case is to allow what should be free to remain free, rather than confine it through a traditional regulatory approach.

The definitions embodied in the political and regulatory system, though, are not as weak in Brazil as they are in Russia. The existing interests still retain a high degree of legitimacy and control over the existing process; markets and commercial laws allow for a better understanding of the value and ownership of existing players than they do in chaotic Russia. But Brazil does have something to learn from the Russian case, especially when it comes to facilitating change in the market for wireless access services. Getting players into the market and allowing them further autonomy creates conditions more likely to lead to successful telecommunications development, especially when telecommunications development is based on wireless access.

None of the other infrastructure development theories referred to in this book would point in such a direction. Those on the side of cultural and technological protection that rightfully point out that the cellular model is draining critical resources from the sector which could be invested in increased access among middle- or low-income families, but the analysis would then create a justification for more government involvement, even though government institutions are the ones that are responsible for and profit from the present arrangements. Further subsidization, in this kind of environment, usually reinforces the existing patterns simply because the corporate institutions that profit from the subsidy are those whose political interests are already closely tied to the state. There is no independent regulator to speak of yet in Brazil, and implementing a regulatory-based model for telecommunications development would only serve to codify existing political and economic relations, rather than open the space that would allow them to evolve.

The theories of the right, though, hold little promise of breaking apart the existing telecommunications entente and replacing it with a constructive alternative. In Brazil, the relationships embodied in the Telebrás system would not permit the easy replacement of political economy by technological economy; perhaps the long and arduous process would eventually reap its rewards, but it is difficult to see exactly how the long-term interests of the country would be served by total liberalization when it would be left in the hands of the institutions that presently control the markets.

It is useful to take a moment and consider the most recent privatization proposal as a starting point for a further discussion of Brazil's future telecommunications development. If the local pole companies are eventually spun off from the Telebrás system, they will also have the opportunity to offer the traditional cellular services while maintaining their existing capabilities in providing local services. Additional providers in Band B will come on line and push penetration rates higher. Perhaps the new Telebrás can be offered the opportunity to develop the strategic liberalization program through wireless access in unused parts of the spectrum, namely the PCS bands of 1.8 to 2.2 GHz.

A new regulator is created, but is weakened by the institutional strength of the technology and telecommunications sector. It is also weakened by the institutional split between local and national telecommunications systems, one similar to the local versus national division in the U.S. but more entrenched because of the weak holding company structure that has been placed on the operating companies. In that kind of environment, the only political goal that the regulator may have the authority to set is that of universal access—every person in Brazil should have the possibility of accessing a telephone, if he or she chooses, through purchasing either a wireless or traditional landline telephone. That sort of generic statement would set the business objective: Companies that push penetration rates higher and increase coverage areas would be rewarded.

That kind of institutional arrangement would provide the flexibility required to meet two divergent goals: rebalancing the authority and resources of the pole companies vis-à-vis the center, while creating a framework for the achievement of a critical national goal. It also takes advantage of the convergence of technologies and forms of wireless access so that Brazil can move forward quickly in establishing a market for both institutions and other participants in the telecommunications sector. But how realistic is it, and how does this one possible evolutionary path differ from others that might appear down the road to a healthier pattern of telecommunications development in Brazil? The final section talks about the categories discussed before in each of the case-study chapters: the level of government regulation, the substance of government policy, and the direction for corporate strategy.

STRATEGIC LIBERALIZATION IN THE BRAZILIAN CONTEXT

In a country with little history of meeting local telephone needs, directing and targeting resources through an appropriately designated institution (or institutions) is critical. In the Russian case, local and city interests are driving the course of telecommunications development. The problem was bringing local efforts together in a fashion that would lead to the construction of a unified communications environment throughout the country. The issue in Brazil is how to shake loose the resources and centralized structure so that it acts a bit more like the Russian case, with local needs being met within a framework that is broadly acceptable to all market participants. Wireless access is an opportunity in Brazil, an opportunity to bridge gaps that appear unbridgeable when liberalization is thought of as primarily an issue to be dealt with in the traditional telecommunications market of wireline services. What does that mean, in practical terms, for the Brazilian case?

The Level of Government Intervention

In the discussions of appropriate levels of government intervention, the recurring theme has been how to remove constraints placed on wireless access providers so as to unleash competitive market that wireless access can support. Those propositions do not have much present meaning in Brazil, simply because the market has only just been launched and the direction forward is still so unclear because of the legal and political difficulties inherent in the privatization process.

But, like the Russian case, there is an opportunity to be informed by some of the failures of the developed world cases and the roadblocks to competition that have been created through patterns of government intervention. Not repeating those patterns will offer Brazil a competitive advantage in the country's future telecommunications development.

License Definitions and Seamless Service Provision: Writing New Telecommunications Law for Brazil

The problem with license definitions, as has been detailed in earlier chapters, is that they restrict the nature of services offered by existing and future wireless service providers. Brazil has adopted a similar pattern of licensing. The regulatory definitions are likely to close off the needed convergence of the wireless access industry and, in an environment where the competitive patterns have not yet been set, there is an opportunity to retain the broader definitions and allow providers to compete in all fields of wireless access.

But that would not be the logical outcome in a country like Brazil. Brazil has, historically, worked to establish a strong central state, either through the military or through state-sponsored development under democratic rule; the tendency will be to define a new telecommunications entente based on centralizing and standardizing. That approach implies strong and fixed license definitions.

The tendency in creating new laws to govern the development of Brazil's telecommunications sector therefore will be to codify existing technological and economic arrangements. Each codification offers potential protection, that is true, but each also can be used by lawyers and politicians in the not too distant future to forestall competitive entrance by new investors, or by incumbent providers wishing to expand the products or services they make available in the marketplace.

The same inherent difficulties exist with regard to seamless service provision. The tendency to rewrite laws that fix a particular kind of telecommunications entente favorable to the interests of existing dominant providers would also imply separating wireline and wireless service provision. There is also the additional possibility that the cellular assets in Band A will be privatized separately from the rest of the Telebrás system. From a public policy perspective, this might be an appropriate way to "break up" Telebrás, but such a policy would likely delay the reintegration and convergence of wireline and wireless access.

Brazil only has to look at the Russian case to see the other risk inherent in constructing the new telecommunications law on that basis: If the law does not fit the needs and interests of the corporate institutions, it may be ignored and the legitimacy of the government and proposed regulatory body would be damaged, perhaps irreparably. Even if the law does fit the times, the flexibility needed for market competition will run contrary to the precepts of government bureaucratic control. For developing world countries attempting to move forward on the course of telecommunications development, vague law is likely to be better than specific law.

The Substance of Government Policy

A number of ideas have surfaced during the discussions of appropriate government policy to support sustainable telecommunications development, some of which may be applicable to the Brazilian context. Given the need to define a new telecommunications entente and the possibility of defining that entente around the expansion and liberalization of the wireless telecommunications market, three items in particular are worth reviewing:

- Future Licensing: PCS and Universal Service Bandwidth;

- Building New Institutional Relationships: International Lending and Investment; and,

- Strengthening the Basis for Technological Innovation.

Licensing: PCS and Universal Service Bandwidth

The first round of licensing of cellular providers has already gone forward before the publication of this book, with the "beauty contest" between various bidders resolved not by competitive auction but by the choice of the local telephone companies. An auction for further spectrum would be a proscription that follows logically from the proscriptions set out, but, considering the nature of the Brazilian case, it might not be advisable. Part of the difficulty for Brazil is to reduce the amount of central involvement in the system and to provide opportunities for competition to emerge where it does not presently exist. An auction, elaborated on the U.S. model, would only serve to further reinforce the authority of the center over the periphery. Additionally, unlike in the United States, the telecommunications sector has already paid the treasury enough during the periods of shrinking capital investment—companies can not afford to see the proceeds of any auction siphoned off into the federal treasury to pay for the national debt.

Opportunities to learn from and benefit from the investment throughout the world in PCS equipment should not be ignored, though. Clearly, a cellular duopoly is not enough competition to lower prices in the United States, so it should not be expected to do so in Brazil. Assuming that the regulatory function is completely decoupled from the telecommunications providers during this round of liberalization and deregulation, it would be best for the established regulatory agency to follow the pattern already set down, licensing two PCS providers in each region with the caveat that neither of the existing cellular players can be licensed for PCS.

The critical innovation would be to set aside a sufficient amount of spectrum (30 to 50 MHz in the 1.8 to 2 GHz band) to take advantage of fixed wireless local loop applications. One fixed wireless license should be granted to the operating company charged with improving the level of penetration within the country. The goal should be to make the entire country accessible to wireless local loop access, with fixed stations in any home that can afford the subscription fee. The government should create an open competition for the technology that would be used to build that network and choose the system that is able to provide the fixed wireless service at the lowest per subscriber cost.

Brazil would establish a new model for universal service, one that is much more likely to be the future of universal service than the concept of wiring each and every home and building, that of "universal service bandwidth." With a wireless access point positioned as an option for every person throughout the country, the potential for universal service would be, in effect, achieved. The question would be the modification of existing economic relationships to increase the resource base of the poorest in the country.

Over time, the fixed wireless system would compete directly with the cellular and PCS systems in place, but that competition would take years to develop. The market should be healthy enough to support a range of providers who can choose to combine different kinds of wireless access to meet the needs of various subscriber groups. At that point, Brazil will have achieved the goal of decentralizing investment and operational capabilities (through cellular and PCS) while simultaneously meeting the needs of national integration and universal service (through fixed wireless local loop).

Building New Institutional Relationships: International Lending and Investment

The government of Brazil will remain responsible for cultivating international lending associations and international investors looking to find uses for capital abroad. The Mexican government's activities leading up to the NAFTA agreement's passage through the U.S. House of Representatives and Senate still stands as one of the best examples of how a government can highlight the favorable portions of its economic program to the long-term benefit of the country as a whole. Brazil will need to take a page from Mexico's book as it moves forward, and strategic liberalization can become a critical part of that effort.

One of the proscriptions of the strategic liberalization strategy was the outsourcing of certain functions, such as billing and marketing where possible and appropriate, to bring down the total cost to each individual customer. The Brazilian government might start a program to construct subsidized institutions that could provide these functions to the cellular systems under development throughout the country. Profitability for these kinds of functions are highly correlated with economies of scale, and secure billing systems in particular are one of the biggest stumbling blocks for foreign investors entering the market for cellular and wireless services.

Investment in this kind of institution can have two ready justifications. To international lending institutions, such as the World Bank and the International Monetary Fund, the construction of such an institution has a directly positive impact on sector development; it brings costs down and improves the opportunities for service provision. For investors, it is an opportunity to ensure that their investment is under the supervision of a billing system that meets with the standards of those in developed countries.

When such an institution takes its place in the market and begins to serve incumbent cellular providers, it will be in a position to provide the same services to the new fixed wireless system that can be constructed to achieve the goals of universal service. The cost to the universal service provider can be set lower than the price for cellular providers until such time as the two forms of wireless access are in direct competition with each other.

This kind of innovative positioning of the country's sector development has the dual advantage of meeting the ideological requirements of foreign capital while meeting a real marketplace need critical for the development of the telecommunications sector. This moves beyond the argument that has traditionally been taken to the financial community from the developing world; this is more than just liberalization and privatization. This is creating institutions that meet both the political needs of development and the economic needs of corporate growth and sector expansion.

Rebuilding the Basis for Technological Innovation

The strategy of import substitution clearly failed Brazil in its efforts to grow and expand local technology and equipment production. The recent downsizing of CPqD is a testament to the inability of a protected, centralized procurement and research system to establish a sustainable structure for telecommunications development. In addition, the crumbling of the centralized authoritarian system now needs to be reflected in the pattern of investment in telecommunications and technology research. Research and development of new technology appropriate to the needs of the Brazilian marketplace needs to be rolled out to the places where it will be used: the local markets.

Brazil can address this issue by borrowing from the pioneer's preference innovation developed by the FCC. Again, although fraught with difficulty, Brazil can put into place a system similar to the one suggested in the chapter on the British system—Brazil can position itself as a testbed for the global telecommunications infrastructure. One example would be the competition for the right to construct (or even operate) the universal service license. There is presently much talk about wireless local loops as low-cost access for telecommunications services throughout the world but, admittedly, there has been little action. Most equipment producers and service providers would much rather serve the customers who can pay more. This kind of contest would force companies throughout the world to focus on the issue of low-cost access in a way that they have yet to do, and few companies would turn away the opportunity to serve one of the largest markets in Latin America, should the opportunity present itself.

Similar kinds of competitions can be structured for critical elements of infrastructure provision. From "beauty contests" to the construction of working trials that can be used in Brazil and throughout the world, there are a number of opportunities to position the country as more than a laggard in a increasingly dynamic region. Such a process would help Brazil regain the initiative in equipment and technological production, an initiative that has been lost since the collapse of the import substitution strategy of the 1970s and 1980s.

The Direction of Corporate Strategy

Comparing the direction of corporate strategy in Brazil to the other countries discussed in this book is difficult; defining a solid ground for corporate strategy in the coming years depends on divining future privatization plans and projecting the emerging basis for Brazil's next telecommunications entente. In the Russian case, where the weakness of the state seems to imply the informalization of the entire telecommunications sector, it is possible to view government policy almost as one would view another corporate institution in the marketplace—sometimes a potential partner, sometimes a real competitor. In Brazil, the tradition of the legitimate state and the strongly held ownership of the Telebrás system makes for a very different situation.

Two areas are at the center of the examination of future corporate strategy: the role of fixed wireless services and the possibilities of using outsourcing and reselling. Each may have a role in the Brazilian context, and are worth reviewing one final time.

Fixed Wireless Services

At the moment, there is no bandwidth available to provide fixed wireless services in Brazil. Although, in theory, such services could be offered in the "A" or "B" cellular bands that are available, it is clear that the cellular model is being implemented and that it will dominate in those bands. Various paging and ESMR licenses have been offered and developed by private concerns, but these do not really provide the kind of opportunity that would attract a private sector institution wishing to offer fixed wireless services.

A company wishing to offer these kinds of services in Brazil has two options. The first is to wait for additional bandwidth to be distributed. This is not likely to happen anytime soon, because policy makers will continue to concentrate on privatization and the development of the new regulatory institution at least until the end of 1998. At that point, with the continued explosion of wireless services around the world even more apparent, the government would have the time to turn attention and resources to creating additional licenses and implementing some sort of scheme to allow fixed wireless loop providers into the marketplace.

The second option is to take the Russian route: focus almost exclusively on lobbying the local and regional political interests in an attempt to ride the wave of local telecommunications autonomy likely to grow in the privatization process. In the Russian case, companies which did this effectively were rewarded with licenses and service opportunities — initially against the wishes of the central government, but eventually with their acquiescence. It can be argued that the success of this strategy is very specific to the Russian case; with the fall of the communist system, each region was invited to grab as much autonomy as it could swallow. No such offer will be forthcoming to the local and regional governments of Brazil. But the history of the country's telecommunications development appears to indicate the importance of this strategy. Telebrás' operations depend on the pole companies, and equipment providers such as Ericsson and NEC have found a great deal of success in selling through the regional entities, even though Telebrás is nominally in charge of infrastructure development and technology choices.

This kind of strategy would be consistent with the pattern of telecommunications development and access competition seen in the U.S. case. The recreation of local autonomy in Brazil, and the push for increased access at the local level, may provide the kind of investment dynamic required to increase levels of penetration among low- and middle-income families throughout Brazil.

To build the market for fixed wireless services in Brazil, the appropriate corporate strategy is one of lobbying and positioning a company for the opportunities which will likely arise in the course of the country's telecommunications development. The success of any strategy will likely depend on matching the company's interests with those at the local and state level. It may be a long wait but, given the potential service territory and market expansion, it will probably be worth the investment.

Outsourcing and Reselling

Existing providers will have the opportunity to reduce costs of cellular services, although probably not to the point where they become a broadly available substitute for wireline services. At the moment, there is no real incentive for the cellular providers, or the pole companies investing in the cellular systems, to change their pricing and infrastructure development strategy. The numbers discussed earlier in the chapter indicate that the cellular model has been fully embraced and that the revenue streams are already substantial enough to declare the initial phase of Brazilian cellular development a rousing success.

Even so, there may be opportunities to partner with emerging companies in a fashion appropriate to corporate behavior in a marketplace duopoly, especially if there is an opportunity to identify regulatory and investment benefits which may accrue to the companies. Outsourcing some of the more expensive portions of the network, including billing and some of the maintenance functions, would be a real opportunity—especially if this was an area where internal lending agencies could be convinced to put their resources as part of the country's development program.

The resale market, though, may prove to be a more viable option in Brazil. In Russia, the history of the country's telecommunications development would appear to discount the importance of resellers, simply because the chaos of the market would not provide a sufficient foundation for long-term resale arrangements. With the emergence of the Brazilian duopoly, the opportunity to develop resale packages will likely develop along the same lines as it did in the United States. Excess capacity on both of the networks, and the tendency to keep prices higher because of the lack of competition in the marketplace, will open the door for lower cost resale offerings. Although the regulatory issues surrounding this kind of strategy are likely to be redefined as part of the privatization process, the opportunity exists to begin establishing private sector presence in this portion of the marketplace, perhaps as the basis for a longer term investment position in the country.

The development of the resale will provide some temporary relief from the imposition of the cellular model, but only until it becomes inconvenient for the incumbent providers. The discussion in the U.S. and U.K. cases made it clear that resale, from the perspective of sustainable infrastructure development, has only a limited role to play in supporting other corporate institutions engaging in facilities-based competition.

CONCLUSIONS

Brazil is certain to continue its role as a bellwether of Latin American development, even though it is lagging behind other countries in the region in the liberalization and privatization of the telecommunications sector. In great part, it has an opportunity to learn from the mistakes of others, but its legacy of central control will be a real burden as it tries to move forward. The tendency will be to reassert central control over an industry that appears to base its business operation on decentralization and establishing horizontal linkages between people and institutions.

The advantage of wireless access, especially as it relates to the Brazilian case, is that its flexibility and potential for targeted kinds of services make it a good opportunity for local and international investment. But Brazil cannot afford to let the cellular and PCS models exist without any competing service opportunity. At the moment, the country is on course to replicate the cellular duopolies that appeared in the U.S. and U.K. during the 1980s, without providing for opportunities in emerging market areas, such as fixed wireless local loop.

The best hope in moving past that stage and accelerating the development of Brazil's telecommunications sector is to free the market for wireless communications. Other companies will have to develop a presence and compete with incumbent systems. But by establishing a new model for cellular service to compete with PCS and cellular systems, Brazil puts into place a foundation for future competition and universal service.

The balancing of local and national interests is a requisite part of the new telecommunications entente that must emerge as Brazil considers the next stage of development. There are areas where privatization will not be enough, and liberalization will have to be carried by the local and regional governments in partnership with corporate institutions wishing to serve the emerging market. With the locals running cellular and the national interests focused on meeting universal service requirements, there may be enough of a compromise for a new telecommunications entente. And compromise is at the foundation of the kind of telecommunications miracle Brazil needs.

ENDNOTES

[1] A great deal of the information included in this chapter appears in "Brazilian Telecommunications in Transition: A New Strategy for Competitiveness," Lee McKnight, W. Russell Neuman, Jose Ferro and Antonio Botelho (Cambridge, MA: Unpublished document on file at the Massachusetts Institute of Technology's Center on Science, Technology, and Policy). Material from that report also appears in Eli Noam's forthcoming book, *Telecommunications in Latin America.*

2   In the Book by Ronald Schneider, *Order and Progress: A Political History of Brazil*, he wrote about the 1922 campaign of the Military Club against the government, the military's role in the Vargas regime, and Vargas' ouster in 1945, and describes the military's involvement as follows:

> The military figures involved in these events have been participants, not just observers, throughout the recurring crises from 1922 on. Hence they felt a sense of responsibility for the outcome of these episodes and second-guessed themselves concerning the sins of omission; a valid comprehension of their perceptions in this regard is at least as essential for understanding the sharp regime change of 1964 as is an analysis of developments within the political system itself. Moreover, when the political system underwent fundamental change, so did the military institution insofar as it functioned as a component of the political system

3   This interpretation of events is strongly stated in the neodependency writings of the past few years, as typified in an article by Anamaria Fadul and Joseph Strabhaar in "Communications, Culture and Informatics in Brazil: The Current Challenges," in Gerald Sussman and John Lent, eds., *Transnational Communications: Wiring the Third World* (London: Sage Publications, 1991). On page 216, they wrote:

> After the 1964 coup, the military thought of communication as an important element also in national security. Two aspects of national security for the military were creating an infrastructure and controlling the image or concept of the nation. The military invested heavily in telecommunications, both in telephony and microwave/satellite links for broadcast networks. The country needed a communications system capable of integrating those elements, to provide a channel through which to portray an image of Brazil that would support the economic model the military wanted. In short, the policy of the time was to provide infrastructure and access, while perpetuating state ideological control, which created major conflicts within opposition political forces.

The position of these authors is, quite likely, overstated. But it would be difficult to deny that the expansion of the telecommunications and media networks closely corresponded with the institutional goals of the newly established military leaders, which was an essential part of the "telecommunications entente" that developed in Brazil and defined telecommunications development for more than twenty years.

4   Capallaro, Jorge Jose V. "Historia da Industria de Equipamentos de Telecomunicacoes no Brasil," in Lins de Barros, Henry British, eds., *Historia da Industria de Telecomunicações* [The History of the Telecommunications Industry], (Historia Geral das Telecomunicações no Brasil. Rio de Janerio: Associacao Brasileira de Telecomunicações).

McKnight and Neuman et. al. argued that the "association between Ericsson and Matec, controlled by the industrial holding Montiro Aranha, was the only one that effectively provided for the transfer of digital technology to a Brazilian associate." ITT, for example, eventually sold out its subsidiary to Brasilianvest along with a transfer agreement for its crossbar technology, but not electronic digital technology. Ericsson has gone on to play a critical role in the manufacture of the Tropico family of exchanges, which became the backbone for the development of Brazil's telecommunications network in the modern age. Ericsson was also partially responsible for the fiber deployment in the São Paulo region, an event which made Brazil the first developing country with a deployed fiber communications capability. As Neuman, McKnight et al. pointed out, though, "the Brazilian telecommunications equipment market is relatively open when compared to those of developed nations such as Japan (with almost 100% domestic equipment), France (86%), or even Italy (65%)."

Conclusion:
Welcome to the Jungle

At the end of his book, *Wireless Access and the Local Telephone Network* (1992), George Calhoun compared the future market for wireless access to a jungle. The abundant energy and a wide variety of wireless "species" and "features" will proliferate, producing biodiversity on the order of a tropical ecosystem. His metaphor is likely to be a very apt description of the future, with mutations and new discoveries beckoning around each hillside and river we encounter.

This book has takes the metaphor one step further. The rainforest is a resource. People draw on that resource for the purposes of development. Given the right conditions, the ecosystem can thrive and provide a richness for those who protect the sustainability of the forest. But what are the right conditions for this particular rainforest known as the telecommunications sector?

It is the human factor—the collective political, economic and social dynamics which shape institutions—that we need to account for in establishing an appropriate policy to ensure the sustainable development of the telecommunications sector. That is why the work of Putnam and the other social scientists we referred to in the first chapter are so important to the ongoing discussion of telecommunications policy. Institutions shape history and history is shaped by institutions. Institutional performance depends on the ability to be shaped by history while directing social and economic energy to goals that are commonly defined as good for a community.

In this book, we defined the common good as the political and economic development of any community or country, bringing resources and capabilities into the hands of those who need them and can use them. At the end of the day, there can be no more compelling common good than this. With all of our theorizing, all of our projections and commentaries, if we are not doing something for the people who should be enfranchised by the institutions of the private and public sector, then the common good is not being served.

So let us take a last look at the forests that we described and the sustainability of institutions in each of those ecosystems. In this final section of comparisons and contrasts, we assess the relative merit of the policy approaches to the telecommunications sector detailed in the first section, and point to the outlines of sustainability in the global ecosystem of our world's information infrastructure.

The Coastal Redwood Forests of the United States

When you travel through the coastal redwood forests of the American northwest, you get the feeling that the trees above you stretch up to infinity. But when you look down, you see the vast diversity of underbrush, the ferns and smaller trees that live and thrive along with the big trees. So many fires have swept through these areas as the seasons change that there are clear signs of renewal, stages of development, and the hardening of the bark of the trees that survived.

The telecommunications ecosystem of the United States has a number of big redwoods like AT&T and the Baby Bells, but also thousands upon thousands of companies that spread out under the shade of these giants. The variety of institutions in this telecommunications ecosystem are a reflection of the wildly diverse history of economic development in the United States. The local and regional emphasis of early telecommunications development marks the landscape; the raging fires of new technologies and economic change have cleared the underbrush on more than one occasion, leaving the ground fertile for the quick rebirth of new forms and species.

If there is any clear trend in the social science research we have reviewed in regard to the United States, it is the fact that any telecommunications development strategy needs to be adapted to the decentralized confusion that shapes the political structure. Local interests need to be at the heart of the development strategy, or else the performance and sustainability of the institutions will remain in doubt.

Our analysis of the telecommunications environment in the United States has revealed a pattern of corporate interest in sustaining the high prices of cellular service, rather than rolling out lower cost alternatives in a more competitive environment. This kind of activity is consistent with oligopolistic competitors in a highly concentrated market; even though there are a number of institutions playing critical roles in the development of the sector, it is clear that the "big trees" are defining the nature of the forest.

The sustainability of the American telecommunications sector is, therefore, threatened by the continued concentration of power and capability within the bigger players in the market. What is needed is a good fire, which removes some of the branches and leaves the underbrush exposed to the light and soil it needs to grow and prosper.

This is where strategic liberalization shows itself to be a policy for sustainable development. By creating a torrent of small- and medium-sized institutions that can compete directly for telecommunications services in the local market, it will be possible to re-establish a pattern of access competition. Wireless access can form the technological foundation for service provision, pushing the existing providers to reinvent themselves yet again and bringing the cost of services down through direct competition.

The analysis also shows where the thinking of the techno-libertarians falls short. It is not a strategy for sustainable development, but rather a scorched earth policy that allows those companies with access to the technology to drive it through the system with little regard for the social consequences. Although there is some evidence that the technology of the digital age supports greater horizontal linkages over vertical linkages, it is also clear that the existing media and bureaucratic structures remain vertically integrated. As such, they have less interest in the underbrush and more interest in their efforts to reach for the heavens.

Sustaining the environment of the coastal redwood ecosystem requires a little revolution. Thomas Jefferson thought it was a good thing for American government, every now and then, and the same holds true for its telecommunications sector. What our research tells us is that the best way to start a revolution that can support sustainable development is to implement a policy of strategic liberalization.

The Oak Trees of the United Kingdom

One of the biggest environmental issues in the U.K. today is the preservation of oak trees. As one of the few indigenous species of trees remaining in Great Britain, and one of the most long-lived of any in the world, concerns have been voiced about the development of certain areas where the oldest oak trees reside. There is also the connection to the U.K.'s spiritual past; the druid sects prevalent in England thought of the oak as having a special vitality and mystical place in the world.

The U.K.'s telecommunications firms are certainly under less threat than the oak trees in certain regions of the country, yet protection is definitely a critical concern reflected in the policy of the century. The traditional strength and stability of British Telecom as a state-run telecommunications provider was valued like "the family silver." To protect it from intrusion, the U.K. established an insulation from competition similar to all of the countries in the developed world.

But the protection was particular to the U.K. case, in great part because of the nature of the public and private sector relationship in the country. The interplay of social, political and economic groups occurred on open ground, providing a vivid context for ideological debate. This relatively open system for discourse infused a sense of urgency and possibility during the 1970s and 1980s, as the Thatcherite regime turned to privatization as the critical linchpin in a strategy of telecommunications development.

But the implementation of a competition policy has been able to take the country only so far. Now, there is a different kind of consensus required, one that involves British engagement with other global telecommunications players, especially in Europe. The U.K. most definitely is one of the global leaders in increasing the level of competition in the telecommunications market but, because of the limited scale of possible services in the country, a framework needs to be set which can be exported—helping the country maintain its competitive advantage in the area of service provision.

Strategic liberalization is a unique opportunity for the U.K., insofar as it ties into and attacks the particular difficulties that have arisen in the market for wireless access services in the country. If wireless access is to be a true, broadly based competitor against wireline access providers, it will be necessary to break through the levels that separate the service provider from the customer.

In this context, the market subsidization philosophy can do little to advance the goals of development through the telecommunications sector. The problems of universal service will be transformed into universal access, and the question will no longer be whether people can get access to basic service, but rather how accessible enhanced services can be in an environment of increased competition. The only way to push that agenda forward is to ensure that investment pushes down the cost of these services so that those who choose to access them get the greatest value from them.

The oak trees of England's telecommunications marketplace are worth preserving, but others need to be grown in the fertile grounds of competition that can be found in the U.K. Strategic liberalization here is a policy of planting and pruning that will help sustain that vibrant orchard.

Russia: The White Birch Trees

Moscow and its environs are filled with small clumps of white birch trees. In parks, near the apartment residences of the urban communities, out in the fieldlands, stretching over the Eurasian plains, the white birches mark the landscape with some color in the trying post-Communist grey of economic and social transition.

The birch is a symbol of vitality and fertility in the Slavic traditions. It is for good reason that Russians take care of their birch trees, but they rarely constitute an entire forest. Good things in Russia come in bits and pieces, apparently disconnected and separated at the surface, even though connected at the roots.

The same can be said for the present stage of Russian economic and political development. The social history of Russia shows a strained relationship between institutions and the environment. Quite often, sustaining institutions in this environment requires a focus on the roots and the locations of the institutions, with the hope that by establishing successes in a number of places, a broader, positive trend will emerge.

The most hopeful scenario for telecommunications development would reflect such a pattern: Separate localities working independently to use the resources at their disposal to construct networks that serve the telecommunications and information needs of their people, yet simultaneously coordinating their efforts to establish an open communications environment. Telecommunications institutions can only be sustainable at the local level, at least for right now—these are the institutions which are reaching the customers and enhancing the resource base of the Russian people.

That's where strategic liberalization has a particularly important role for Russia's telecommunications development. The small- to medium-sized institutions that have begun to grow in the regions of Russia will not be able to sustain developed-world levels of investment and institutional support. They will have to respond to difficult weather conditions, and certainly there will be a few cold winters in the coming years to test the capabilities and resolve of the companies that develop. The use of wireless technologies among service providers, and the specific policies that Russia could implement to improve investment opportunities in wireless access, may make the difference between a healthy telecommunications sector and one that dies stillborn in the next century.

This is a condition in which the perspective of those in the regulatory mainstream does not apply. Those who focus almost exclusively on the nature of the regulatory institutions and the macroeconomic policies are missing the forest for the trees; it is the micropolitical and microeconomic impact that is most critical in the Russian case. If the regions do not produce strong shoots of growth and development, nothing on the national level will be sustainable and simultaneously achieve the objectives of development we have set out as our criterion for performance.

With a policy of strategic liberalization in place, Russia will be able to sustain regional telecommunications development by providing an appropriate technological foundation and a sufficient financial base. Perhaps, the birch trees will be able to spread and grow even stronger, forming a chain of vitality that lifts Russia up and into the future.

Brazil: The Mahogany Trees of the Amazon Rainforest

One of the most valuable resources of the Brazilian Amazon rainforest is the mahogany trees. They are often the focus for private sector investments in the logging industry; those who do not follow sustainable harvesting policies often rip down large portions of the forest, extracting from the resources only the mahogany and other valuable trees while discarding the rest.

In many ways, the procedural entente of the military period in the history of Brazil's telecommunications development is like this logging policy: The valuable resources have been extracted at the expense of the whole forest. Money and resources drawn out of the Telebrás holding company during the 1970s and 1980s diminished some of the positive effects of centralization, while exacerbating some of the inherent difficulties in this kind of development exercise.

The history of Brazil tells us that the effective implementation of public policy requires a particular kind of procedural consensus among key actors in the public and private sector, including interests groups like the military and labor unions. The telecommunications development of the country will depend on establishing a framework for common action and investment so that the country can move forward through the privatization stage and into a new environment more amenable to sustainable telecommunications development.

Strategic liberalization is appropriate for the Brazilian case insofar at it begins with an asset whose patterns of ownership and development have yet to be established. The fact that wireless access is a relatively new phenomenon gives it an advantage over other kinds of telecommunications development: The nature and purpose of service provision remains an open question, and new kinds of institutions can be built to support the development of wireless access.

The policies laid out in this book can provide the beginnings for an ongoing discussion about the role of wireless access in an evolving marketplace, offering enough support for a very different kind of institution in the Brazilian telecommunications marketplace. Service institutions based on wireless access will be able to make competitive inroads where other providers cannot, while simultaneously establishing a new focus for investment and infrastructure construction critical to the modern age.

Many policy analysts in the cultural and technological protection camp would disagree with one of the fundamental principles of strategic liberalization as a model of telecommunications development. They would claim that the openness to market investment and resource allocation puts at risk some of the country's most valuable assets, transferring the role of the unscrupulous logger from the state to the private sector. Such an argument, though, ignores the value that would be brought into the environment if an open marketplace were established. There are certainly difficulties inherent in the opening of any markets so closely protected by government regulation, but the benefits in terms of technological diffusion and innovation are likely to outweigh the costs.

Sustaining the value of the whole jungle is the critical point of a strategic liberalization strategy. By focusing the energy and resources of the world on the national market of Brazil, the country will have a much better opportunity to leverage the country's telecommunications development to meet the overall objectives of economic growth and political modernization.

Comparing Case Studies: The Grounds for Common Action

One of the significant themes in the opening section of this book concerned the need for common values and themes for the promotion of telecommunications development as a critical portion of achieving increased rates of economic growth, political participation, and social modernization. The examination of telecommunications development in the four countries discussed makes clear a basic assumption of this book and the literature of the social sciences as a whole with regard to the connection between social values, government policy, and corporate activity: Shared values, at the end of the day, are what ensure the sustainability of institutions.

The broader values for corporate and public policy identified in the first chapter were at the center of the discussion: institutional sustainability and increased penetration of services. The case studies brought forth a host of others that are connected to these two, and there are a few that are worth focusing on in detail as part of the broader discussion of creating a sustainable environment for telecommunications investment.

The first value is the need for a local, "bottom-up" approach to telecommunications development. Even in a time of increasing globalization in telecommunications services, and the perception of competition in long-distance and value added services, local penetration of telecommunications infrastructure needs to be emphasized. As these countries struggle with the harmonization of local and national interests, decisions will have to concern local infrastructures. But that does not mean local town councils and other government actors should hold sway—it is local infrastructures and local access that form the basis of sustainable policy and investment.

The United States and Russia have direct historical experience with the "ground up" development of services, although the recent past of Russia's chaotic telecommunications development pattern looks significantly different from the American examples from a century ago. The opportunity to use "access competition" in a country like Brazil represents a unique opportunity, but also requires a certain kind of consensus among regional powers to assume responsibility for driving telecommunications development and wrest it away from the hands of the government and Telebrás. History has shown that Brazilian provincial authorities are more than able to do so, but the goal, again, should not be to increase the power of local government, but rather put capabilities into the hands of private service providers that want to build new and competitive infrastructures. By serving the needs of local and regional communities, and allowing telecommunications companies to compete for customers by providing local services, access competition as a model represents the value of "bottom-up" telecommunications development.

But for the United Kingdom, this is a more difficult proposition. A country with less of a scale than the other three case studies, the United Kingdom can only sustain so many providers in so many localities of interest to private sector companies. Already, there is a consolidation underway that will bring together various cable, cellular, PCS, and traditional wireline telephony companies. How does that development square with the value of bottom-up telecommunications development?

Even for countries with smaller potential to offer economies of scale and sustain the number of providers of the United States or other larger countries, there is still an opportunity to push local investment and telecommunications development through strategic liberalization. Low-cost providers, either in the field of wireless cable, modified Telepoint services, or satellite providers, will have a critical role to play in competing for customers—as long as regulators and government officials recognize the strategic value of these investments and lower the barriers to entry that would prevent this development.

But, perhaps more critically, the internationalization of telecommunications markets makes the small country difficulties in adopting this kind of value as the basis for public policy somewhat more possible. Most telecommunications providers will be looking to serve Europe as a geographical region, rather than just the United Kingdom alone. Because some of the most advanced telecommunications technologies in the world will be in place in the more competitive United Kingdom market, the tendency will be to use the U.K. as a take-off point for European ventures—especially if the U.K. can stay ahead of the regulatory curve. But the difficulty will be to ensure that companies invest in local infrastructures, not just the long-distance and international connections which serve the business community. That is why adopting the bottom up model, as a principle for corporate and public policy, is critical even for a smaller country like the U.K.

A second value that has lingered in the background since the first chapter also deserves to be brought back into the forefront: appropriateness. The literature of "appropriate technology" offered some direction as to the appropriate direction for modernization, but this discussion has taken the analysis a bit farther. What is clear from the analysis of the case study countries is that the telecommunications infrastructure is a reflection of, at least in part, the dominant political and economic patterns of the country discussed. That means, no matter what portion of the political spectrum one has as the starting point of liberalization and the transition to a more competitive marketplace, there is a role for political economy to play in contributing to the overall direction of public policy and corporate strategy.

For the United States, the defining characteristics of the local and national authority and jurisdiction have shaped the past of telecommunications services. Any appropriate policy solution or corporate strategy will be grounded in those present realities; there is no clean slate from which to begin, and even the most powerful technologies can not wipe clean the centuries-old pattern of development in the United States. The need to bridge the gap will define what is appropriate in the American context and what is not.

The context for development in the U.K. is different. A certain kind of policy consensus needs to develop, but it is a consensus built on agreement with a different set of constituencies through a pluralistic, more unitary process than in the United States. The appropriate solutions will be focused around Britain's unique ability, at least in the European sphere, to maintain an open consultative process in the development of social and corporate policy. Institutions building infrastructures in the U.K. need to connect with that tradition and use it to anchor the company's service presence, corporate reputation, and infrastructure capabilities.

Brazil and Russia, as developing countries making the transition to a new kind of market economy, have a restricted elite who have access to the great majority of resources available for telecommunications development. Often, multinational corporations and international lending agencies try to circumvent these elites because they are perceived to be less than trustworthy (and, in many cases, they have kept most of their money offshore and do not wish to invest it in their own countries for fear of the potential for loss). Somehow, these investors and political actors need to be engaged by a broader argument than their own self-interest. In the Brazilian case, history has shown that, on the broader issues, consensus does arise, enough to carry it from one stage of history to another, albeit with less progress than might be possible in a more centralized, "orderly" society. Russia likely will face the same constraints; without a specific consensus about appropriate development within Russia, though, there can be little hope for systematic progress. The inability of weak states in the developing world provides fewer resources to remove barriers on the way to better development strategies for the telecommunications sector.

For all of these countries, though, specific kinds of arguments need to be made that link corporate and public interests together. Those arguments need to be driven by a process that is appropriate to each country and aligned with the social and economic development of the country, but that also takes advantage of the specific circumstances and kinds of "ententes" which are important to the people of the country. Appropriate combinations of process and development strategy can provide a common framework for each nation to move forward in aligning the goals of telecommunications development with those of sustainability, increased service penetration, social and economic modernization and political participation.

TELECOMMUNICATIONS BIODIVERSITY AND THE SUSTAINABILITY OF THE GLOBAL INFORMATION INFRASTRUCTURE

As always, an examination of case studies and the specific learnings that can be derived from them leaves an analyst looking at the trees, not the forest. It is time to move back even further from the individual examples and ask what is perhaps the most compelling issue for telecommunications development over the course of the next generation: How will competition affect the evolution of the global telecommunications marketplace? And, more critically for our purposes, what does the policy of strategic liberalization mean for this global market?

Change and Continuity in Telecommunications Development

In each of the environments examined, one of the constants has been the immense difficulties for corporate institutions as they struggle with technological change. The range of adaptations is enormous, and the strategies undertaken to ensure institutional sustainability during this increasingly unsure environment are as numerous as the companies involved in the industry. Telecommunications development policies will have to reinforce institutional flexibility by tying performance to alterations in the social and technological environment.

Surprisingly, we found many similarities in the transformation of private and public sector institutions in the telecommunications industry. We identified a kind of path dependency, where certain kinds of telecommunications development have become institutionally entrenched, even while social and technological facts are conspiring to make them irrelevant. The dominant institutions of the telecommunications sector have benefited from the existing arrangements.

Nevertheless, even with these advantages, the institutions are revolutionizing themselves. The privatizations in the U.K. and throughout the world are one such example; corporate downsizing in all of the telecommunications companies in the U.S. is another. Brazil is moving toward privatization and more regional influence. The political and economic developments in Russia are supporting the establishment of strong local centers for the most viable institutions of the country's telecommunications sector. These changes are occurring because of the force of history and the promise of what lies ahead. Institutions are changing to take advantage of the opportunities of technological and social change as they appear.

Many of the similarities in institutional transformation can be ascribed to the global nature of the sources of change. Awareness of new technologies is quickly diffused among the dominant players in the telecommunications industry. The end of the cold war and the acceptance of the free-market models for development radically transformed the intellectual platform on which policy decisions are made. An increasingly information-hungry market is asking for more access, cheaper, faster, and better. The institutions responsible for providing and regulating the provision of services will have to keep up with the demands of technologies and marketplaces.

But the differences in institutional transformation are substantial. These differences can be ascribed to political and social factors individual to each environment, which can cause variance between existing institutions and possible paths for developmental change. Perhaps just as significantly, the pace of technological change is uneven, and it is doubtful that any one institution can incorporate all of the technological innovations of the telecommunications, information and computer industries.

These patterns of continuity and change are familiar to social scientists examining the sources of institutional change in a comparative context. For those who attempt to define actionable and effective policy through such an analytic framework, the key has always been institutional flexibility. As we discussed in some depth in our first chapter, the new opportunities for participation and modernization require a new kind of public policy for all sectors of the economy, especially those directly involved in infrastructure provision. By establishing a policy that allows for institutional flexibility, participation by a broad variety of groups and interests becomes more likely—and more sustainable.

For the telecommunications industry, flexibility should result in a wide diversity of companies, serving a range of markets and peoples throughout the world. We need a global telecommunications infrastructure policy to promote the diversity of telecommunications institutions. By ensuring that there are a variety of infrastructures, corporate strategies, and business approaches, we can best ensure the health and well-being of this increasingly rich environment.

Global Relationships and Telecommunications Development

This need for flexibility takes on particular importance as we begin to gaze into our hazy crystal ball and recognize the consequences of any competitive policy. As the world begins to move beyond nation-states, and social institutions reflect communities defined more by interest and affinity than national borders, the importance of sustaining a global information infrastructure cannot be understated. As a link that will bring people together (and divide people from each other), the architecture of the world's communications infrastructure will define, in great part, how our global economy and political system works.

These individual telecommunications ecosystems are learning to coexist with each other, uneasily at first. But, like today's world of commerce, they will be increasingly interdependent and "infect" each other with their particular traits and patterns. As always, the dynamics of developed world systems are likely to weigh heavily on the developing world as it struggles to make do with the limited resources it has. The need for balance has never been more apparent in the global environment of telecommunications development.

Even during this period of globalization, localities will become more and more critical as technology (in the hands of some but not all) is used as a device to reinforce local interests over what have been dominant centers of authority. This kind of development can be seen in the modern corporation, where decision making is increasingly decentralized, and even in the public sector, where local political institutions (and, in some cases, military institutions) are beginning to reassert their power vis-à-vis the nation-state that has governed them for so long. This process is being defined by the communications networks being put in place today, and the institutions that arise will be reinforced by the centers of gravity that emerge from the nexus of technology and institutional power.

But the existing institutions of the global information infrastructure are national in scope and purpose. As has been clear from the preceding discussion, nation-states have acquired the control technology of the telephone and the telegraph and put it to use in the purposes of aggregating the authority of the center. British Telecom, Telebrás, Rosstelekom, even AT&T, though it was in private hands, were very much in the service of the nation-states that protected them.

This difficult balance between local needs and international markets adds complexity to what is already a difficult issue: How can the diversity of telecommunications providers be sustained on a global scale? This is very important for developing countries with weaker institutions in the telecommunications sector. Certainly, the national telecommunications providers of most developed world countries will be able to address local needs, but if diversity is only diversity for the rich countries, then too much will be lost in the process.

STRATEGIC LIBERALIZATION: SUSTAINING AND DIRECTING TELECOMMUNICATIONS DIVERSITY

Wireless access is not just another opportunity; it is much more revolutionary for the global context than many perceive. The opportunities for political and economic participation are radically changed through wireless access. Through a program of strategic liberalization, multiple service architectures for the provision of services can be established and sustained. Over time, those platforms will become global, reaching out to people and offering a new kind of seamless communications.

The introduction of new technologies will certainly have an impact on economic modernization, changing social and political organization in course. Wireless access will lend a different character to modernization, one that is more likely to address some of the critical problems of investment and access better than other kinds of telecommunications technologies. Public policy needs to focus on directing the energies and investment of those who will be building the infrastructure and institutions of the future; by providing general parameters in the form of regulation and rewards for those who are willing to risk, policy institutions can bring some formidable resources to bear in defining the future of the global information infrastructure.

But for strategic liberalization to become a protector of the global diversity in the telecommunications sector, there will need to be a global consensus on a new model for the provision of wireless access services. A continued emphasis on privatization without liberalization will take its toll on the viability and sustainability of new institutions employing wireless access for competitive advantage. The expansive growth rates of the developed world are already beginning to slow as service reaches saturation among high-end users, such as corporate executives and upper-income families. New species will only evolve if the environment allows them to grow and prosper.

This is where global relationships will be critical to ensuring that wireless access and telecommunications investment are linked to the goals of national development, not just corporate profits. Institutions like the World Bank and the International Telecommunications Union have to engage the institutions of the public and private sector in this issue, establishing new service models for wireless access and other forms of telecommunications service. Without strong advocates for strategic liberalization in the global telecommunications marketplace, the implementation of this model in individual countries will be limited.

Establishing this advocacy role will be perhaps the most difficult proscription of the strategic liberalization policy we have outlined, simply because it will not be popular with a number of the constituencies within the public and private sector institutions we have put under the analytical microscope.

The regulators and politicians will look at the evident inconveniences. In the past, the telecommunications sector has been marked by its stability and rock-solid financial and political footing. That will change as new institutions offering wireless services begin to compete directly with the landline platform in place today. The opening round has begun in the U.K., and as PCS licensees come on line in the United States, the push to invade the local loop will heat up. In Brazil and Russia, wireless is already seen as a bypass, and further development of the sector (especially through fixed wireless local loop applications) is likely to explode the existing wireline monopoly.

Many corporate institutions have co-opted wireless access to deliver a different kind of product to consumers, based on the cellular service model. As we made clear, wireless access can be much more than a phone on the move, but, for many of these institutions, such a service model would eat into their margins, already under pressure because of increased competition for landline access. Many will look at this policy proscription and not recognize the need to invest in long-term profitability rather than short-term returns.

The environment is likely to be far from stable. Much in the same way that an ecosystem goes through cycles of creation and destruction, so too will the global telecommunications sector. Companies will collapse. They will go out of business and cease to serve customers. Such a scenario is possible in any of the case study countries we have examined. An RBOC falters under the weight of competition as wireless access companies (especially AT&T in the opening rounds of the battle) cut prices to the bone to establish a firm subscriber base. Mercury's wireless operation folds because of increased competition and its wireline operation languishes until the company loses its relevance and others take its place from a resurgent European continent. Rostelekom is pushed out of the business of providing service because privatization fails yet again and local phone companies no longer have to deal with the company. Embratel loses its market share as pole companies conspire against the center. All of these are serious possibilities and need to be considered as the future course of telecommunications development is charted.

Wireless service providers are going to be the most fragile institutions in the opening rounds of competition, simply because the barriers to entry can be dropped quickly and new networks can come on line just as fast. Because of the success of auctions, governments are likely to put more and more spectrum up for grabs, adding the number of players until the market for spectrum itself goes bust.

Are we prepared for this? In a truly competitive market, where a variety of players offer services through a variety of platforms, nothing would happen in terms of service. Another provider would pick up the subscribers, or a new entrant would attempt to take the opportunity to gain a market position. We will have to live with a global information infrastructure that fails, on occasion. Like the computers that will manage the networks, parts of it will have to be "rebooted," but the loss of one institution will create new opportunities in the process of "creative destruction."

Wireless access, in very critical ways, is much more amenable to creative destruction. Its scalability, price, and flexibility give it advantages over the lumbering wireline networks of each of the countries examined. When combined with specific policies, we have seen that wireless access presents the best hope for the development of stronger national information infrastructures through competition. On a global scale, wireless access is likely to provide international players more flexibility and capability to compete and meet customer needs.

Closing Thoughts

Of the emerging opportunities for corporate and public managers in the telecommunications sector, wireless access directly and positively affects the ability of communities to harness the capabilities they need to move forward on the path to development. And, considering the future trends of competition and globalization in the telecommunications industry, the problem of access is best solved by a technology that gives institutions the flexibility to serve, to profit, and to fail, if need be, through the process of creative destruction that marks modern development.

If the future of wireless access and the global telecommunications sector is a jungle, then strategic liberalization is a map to lead us through that jungle, allowing us to plant seeds along the way to further its growth and allow us to harness the rich value of the jungle's diversity. Welcome to the jungle indeed; a global infrastructure of companies rising and falling, of species mutating and altering before our very eyes, of people struggling with the new opportunities for development and moving on with their lives. The jungle of the telecommunications future may not be as bright and magical as we would initially hope. But at least we can try to ensure that the goals of development are securely tied to the growth and expansion of each country's telecommunications infrastructure.

References

Abramson, J. B., Arterton F. C., & Orren, G. R. (1988). *The electronic commonwealth; The impact of new media technologies on democratic politics.* New York: Basic Books.

Abu-Lugod, J. (1989). *Before european hegemony.* Oxford: Oxford University Press.

Adas, M. (1989). *Machines as the measure of men; Science, technology and ideologies of western dominance.* Ithaica: Cornell University Press.

Alder, E. (1987). *The power of ideology: The quest for technological autonomy in Argentina and Brazil.* Berkeley, CA: University of California Press.

Almond, G. (September, 1988). The return to the state. *American Political Science Review*, 82, 120-143.

Almond, G. (1990). *A discipline divided: Schools and sects in political science* Newbery Park, CA: Sage Publications.

Anderson, B. (1983). *Imagined communities.* New York, NY: Verso Publications.

Arora, S. K., & Lasswell, H. D. (1969). *Political communication; The public language of political elites in India and the United States.* New York: Holt Rinehart & Winston Inc.

Baldwin, F. G. C. (1938). *The history of the telephone in the United Kingdom.* London: Chapman & Hall.

Balston, D. M. & Macario, R. C. V. (eds). (1992). *Cellular radio systems.* Norwood, MA: Artech House.

Baker, W. J. (1970). *The history of the Marconi company.* London: Methuen.

Barber, B. R. (1984). *Strong democracy; Participatory politics for a new age.* Berkeley, CA: University of California Press

Baumol, W. J., Panzar, J. C. & Willig, R. D. (1988). *Contestable markets and the theory of industrial structure.* New York: Harcourt Brace.

Bell, D. (1973). *The coming of post industrial society.* New York: Basic Books.

Benjamin, G. (ed). (1982). *The communications revolution in politics: Proceedings of the academy of political science.* New York, NY: The Academy of Political Science.

Beniger, J. R. (1986). *The control revolution.* Cambridge, MA: Harvard University Press.

Blaug, M. (1986). *Economic theory in retrospect.* Cambridge: Cambridge University Press

Blumler, J., & McQuail, D. (1969). *Television in politics.* Chicago, IL: University of Chicago Press.

Branscomb, A. W. (ed). (1986). *Toward a law of global communications networks.* New York: Longman.

Brock, G. W. (1981). *The telecommunications industry.* Cambridge, MA: Harvard University Press.

Brooks, F. P. (1982). *the mythical man-month; essays on software engineering*. Reading, MA: Addison-Wesley.

Brotman, S. N. (ed). (1987). *The telecommunications deregulation sourcebook*. Boston, MA: Artech House.

Brown, R. D. (1989). *Knowledge is power; The diffusion of information in early America, 1700-1865*. New York: Oxford University Press.

Burns, E. B. (1993). *A history of Brazil*. New York: Cambridge University Press.

Brzezinski, Z. (1990). *The grand failure*. New York: Macmillan.

Calhoun, G. (1988). *Digital wireless radio*. Norwood, MA: Artech House.

Calhoun, G. (1992). *Wireless access and the local telephone network*. Norwood, MA: Artech House.

Campbell, J. (1982). *Grammatical man*. New York: Simon & Schuster.

Campbell, R. W. (1988). *The Soviet telecommunications system*. Indianapolis, IN: The Hudson Institute.

Cardoso, F. H. & Faletto, E. (1979). *Dependency and development in Latin America*. Berkeley: University of California Press.

Casado, F. (June, 1996) Substitution effect of mobile telephones on fixed telephony. *Presentation to the Eleventh Biennial Conference, International Telecommunications Society*.

Chandler, A. (1990). *Scale and scope; the dynamics of industrial capitalism*. Cambridge, MA: Harvard University Press.

Clapham, C. (1985). *Third world politics; An introduction*. Madison, WI: The University of Wisconsin Press.

Clapham, M. (1957). *Printing; a history of technology*. New York: Oxford University Press.

Coopers & Lybrand. (1992). *Speeches from the international telecommunications group conferences*. Philadelphia, PA: Coopers & Lybrand.

Coopers & Lybrand. (1993). *Speeches from the international telecommunications group conferences*. Philadelphia, PA: Coopers & Lybrand.

Coopers & Lybrand. (1994). *Speeches from the international telecommunications group conferences*. Philadelphia, PA: Coopers & Lybrand.

Copeland, T. (1994). *Valuation: Measuring and managing the value of companies*. New York: Wiley.

Crandall, R. (Summer, 1992). Regulating communications: Creating monopoly while protecting us from it. *The Brookings Review, vol 34*.

Crandall, R., & Flamm, K. (eds). (1989). *Changing the Rules: Technological Change, International Competition and Regulation in Communications*. Washington, DC: The Brookings Institution.

Crankshaw, Edward. (1976). *The shadow of the winter palace*. New York, NY: Viking Press.

Cronin, F. J. (1993). *Pennsylvania telecommunications infrastructure study*. Harrisburg, PA: Public Utility Commission, 1993.

Cronin, F. J., Colleran, E. K., Herbert, P. L., Lewitsky, S. (August, 1993). Telecommunications and growth: The contribution of telecommunications infrastructure investment to Aggregate and Sectoral Productivity. *Telecommunications Policy*, 17 (9), 677-690.

Dahl, R. A. (1991). *Modern political analysis*. Englewood Cliffs, NJ: Prentice-Hall.

Davies, A. (1994). *Telecommunications and politics*. New York: St. Martin's Press.

Derry, T. K., & Williams, T. I. (1960). *A short history of technology*. Oxford: Oxford University Press.

Deutsch, K. (1963). *The nerves of government*. London: The Free Press of Glencoe.

Dinsmoor, J. (1990). *Brazil: Responses to debt crisis, impact on savings, investment and growth*. Washington, DC: Inter-American Development Bank.

Director, M. D. (1992). *Restructuring and expanding national telecommunications markets*. Washington, DC: The Annenberg Washington Program.

Donaldson, Lufkin & Jenrette. (1993). *1993 cellular communications industry report* New York: Donaldson, Lufkin, & Jenrette.

Donaldson, Lufkin & Jenrette. (1996). *1996 wireless communications industry report* New York: Donaldson, Lufkin, & Jenrette.

Duch, R. (1991). *Privatizing the economy: Telecommunications policy in comparative perspective*. Ann Arbor: University of Michigan Press.

Dunkel, T. (November, 1994). Staking a claim on the information highway. *Working Woman*.

Durkheim, E. (1981). *The elementary forms of the religious life*. New York: The Macmillian Press.

Dutton, W. H., Blumler, J. & Kraemer, K. (eds). (1987). *Wired cities, shaping the future of communication*. Boston, MA: G. K. Hall.

Edelman, M. (1967). *The symbolic uses of politics*. Urbana, IL: The University of Illinois Press.

Egan, B., & Wildman, S. (October, 1994). Funding the public telecommunications infrastructure. *Presentation to the 22nd Annual Telecommunications Policy Research Conference*.

Eisenstien, E. L. (1979). *The printing press as an agent of social change*. Cambridge: Cambridge University Press.

Eisenstien, E. L. (1983). *The printing revolution in early modern Europe*. Cambridge: Cambridge University Press.

Elias, N. (1978). *The civilizing process*. New York: Urizen Books.

Ellul, J. (1990). *The technological bluff*. Grand Rapids, MI: William B. Erdmans Company.

Escutia, E. P. (June, 1996). The financing process of mobile networks. *Presentation to the Eleventh Biennial Conference, International Telecommunications Society*.

Febvre, L., & Martin, H. (1976). *The coming of the book; The impact of printing 1450-1800*. New York: Verso Publications.

Francis, J. (1993). *The politics of regulation: A comparative perspective*. London: Blackwell.

Frederick, H. H. (1993). *Global communication & international relations*. Belmont, CA: Wadsworth.

Frydman, R., & Rapaczynski, A. (1994). *Privatization in Eastern Europe: Is the state withering away?* New York: Central European University Press.

Fukuyama, F. (1992). *The end of history and the last man*. New York: Maxwell Macmillan International.

Gabel, D. (September, 1994). The early competitive era in telephone communications, 1893-1920. *The Journal of Regulatory Economics*.

Gabel, D., & Kennet, D. M. (September, 1994). Economies of scope in the local telephone exchange market. *The Journal of Regulatory Economics*.

Ganley, O. H. & Ganley, G. D. (1982). *To Inform or control? The new communications networks*. New York: McGraw-Hill.

Garrison, M. (1988) *Four case studies of structural alterations in the telecommunications industry*. Washington, DC: Annenberg Washington Program in Communications Studies.

Geertz, C. (1963). *The interpretation of cultures.* New York: Basic Books.

Geller, H. (1991). *Fiber optics: An opportunity for a new policy.* Washington DC: The Annenberg Washington Program.

Gellner, E. (1985). *Relativism and the social sciences.* Cambridge: The University of Cambridge Press.

Gellner, E. (1988). *Plough, sword and book.* Chicago, IL: The University of Chicago Press.

Gerbner, G. (1977). *Mass media politics in changing cultures.* New York: John Wiley.

Gilder, G. (August, 1994). Auctioning the airwaves. *Forbes ASAP.*

Gillis, M., Perkins, D., Roemer M., & Snodgrass, D. (1987). *Economics of development.* New York: W.W. Norton.

Gillick, D. (February, 1991) Telecommunications policy in the UK. *Telecommunications policy.* 15 (2), 140-155.

Goldman, S. L., Nagel, R. N., & Preiss, K. (1995). *Agile competitors and virtual organizations: Strategies for enriching the customer.* New York: Van Nostrand Reinhold.

Goody, J. (1977). *The domestication of the savage mind.* Cambridge: Cambridge University Press.

Goulet, D. (1977). *The uncertain promise, value conflicts in technology transfer.* New York: Overseas Development Council.

Gustafson, T. (1989). *Crisis amid plenty: The politics of soviet energy under Brezhnev and Gorbachev.* Princeton, NJ: Princeton University Press

HRN. (June, 1996). *Public affairs in the 21st century: A speech by Diana Shayon to the public affairs council.* Washington, DC: Public Affairs Council.

Habermas, J. (1979). *Communication and the evolution of society.* Boston: Beacon Press.

Habermas, J. (1984). *The theory of communicative action.* Boston, MA: Beacon Press.

Hachten, W. (1983). *The growth of media in the Third World.* Iowa: Iowa State University Press.

Hanke, S.H., ed. (1987). *Privatization and development.* San Francisco, CA: International Center for Economic Growth.

Hanson, J. & Narula, U. (eds). (1990). *New communications technologies in developing countries.* Hillsdale, NJ: Lawrence Erlbaum Associates.

Hamelink, C. J. (1983). *Cultural autonomy in global communications.* New York: Longman.

Harel, D. (1987). *Algorithmics; the spirit of computing.* Reading, MA: Addison-Wesley.

Hart, J. A. (1988). The politics of global competition in the telecommunications industry. *The Information Society.*

Harrison, P. (1990). *Inside the third world.* London: Penguin Books.

Hazlett, T. W. (September, 1995). Assigning property rights to radio spectrum users: Why did FCC license auctions take 67 years? *Presentation to the 23rd Annual Telecommunications Policy Research Conference.* Solomons Island, MD.

Heldman, R.K. (1992) *Global communications; Layered networks, layered services.* New York: McGraw-Hill.

Hernstein, R. J., & Murray, C. (1994). *The bell curve: Intelligence and class structure in American life.* New York: The Free Press.

Hills, J. (1986). *Deregulating telecoms: Competition and control in the United States, Japan and Britain.* London: Quorum.

Hills, J. (April, 1993). Back to the future: Britian's 19th century telecommunications policy. *Telecommunications Policy.* 17 (3), 256-270.

Hirshman, A. O. (1970). *Exit, voice, and loyalty.* Cambridge, MA: Harvard University Press.

Horwitz, R. B. (1989). *The irony of regulatory reform: The deregulation of American telecommunications.* New York: Oxford University Press.

Huber, P., Kellogg, M. K., & Thorne, J. (1993). *The geodesic network II.* Washington, DC: The Geodesic Company.

Huber, P. (1987). *The geodesic network: Report on competition in thet telephone industry.* Washington, DC: U.S. Department of Justice.

Hudson, H. (1984). *When telephones reach the village.* Norwood, NJ: Ablex.

Hudson, H. (1990). *Communication satellites: Their development and impact.* New York: Free Press.

Hudson, H. (September, 1994). Access to telecommunications in the developing world: Ten years after the Maitland Report. *Presentation to the Twenty-Second Annual Telecommunications Policy Research Conference.* Solomons Island, MD.

Huntington, S. (1968). *Political order in changing societies.* New Haven, CT: Yale University Press.

Innes, J.E. (1975) *Knowledge and public policy; The search for meaningful indicators.* New Brunswick: Transaction Publishers.

Innis, H. (1954). *The bias of communication.* Toronto: University of Toronto Press.

Innis, H. (1972). *Empire and communications.* Toronto: University of Toronto Press.

International Institute of Communication. (1994). *Closing the telecommunications gap.* Geneva: International Telecommunications Union World Telecommunications Development Conference Study Paper.

International Telecommunications Union. (1994). *World telecommunications development report.* Geneva: ITU Publications.

International Telecommunications Union. (1995). *World telecommunications development report.* Geneva: ITU Publications.

International Telecommunications Union. (1996). *World telecommunications development report.* Geneva: ITU Publications.

Jerpersen, J., & Fitz-Randolph, J. (1981). *Mercury's web: The story of telecommunications.* New York, NY: Athaneum.

Johnson, D., & Macomber, B.K. (August, 1994). Laying a sound business foundation. *Private Cable & Wireless.*

Jones, L. P., Tandon, P., & Vogelsand, I. (1990). *Selling public enterprises.* Cambridge, MA: The MIT Press.

Jussawalla, M. (ed). (1993). *Global telecommunications policies: The challenge of change.* Westport, MA: Greenwood.

Kagan, P. (1994). *Cellular telephone atlas.* Carmel, CA: Paul Kagan Associates.

Kagan, P. (1995). *Cellular telephone atlas.* Carmel, CA: Paul Kagan Associates.

Katz, E. & Wedell, G. (1977). *Broadcasting in the third world; promise and performance.* Cambridge, MA: Harvard University Press.

Katz, J. E. (June, 1996) Social consequences of wireless communications: A selective analysis of residential and business sectors in the United States. *Presentation to the Eleventh Biennial Conference, International Telecommunications Society.* Seville, Spain.

Katz, R. (1988). *The information society: An international perspective.* New York: Praeger.

Kazachkov, M., Knight P., & Regli, B. (July, 1996). Using distance learning to facilitate the transformation of the regulatory, business and social environment in Russia. *Presentation to the Second International Conference on Distance Education in Russia.* Moscow, Russia.

Kellerman, A. (December, 1990). International telecommunications around the world; flow analysis. *Telecommunications Policy.* 14 (10), 742-765.

Keith, H. H. & Hayes, R. A. (eds). (1976). *Perspectives on armed politics in Brazil.* Tempe, AZ: Arizona State University Press.

Kuhn, T. (1961). *The structure of scientific revolutions.* Chicago, IL: University of Chicago Press.

Kurisaki, Y. (June, 1993). Globalization or regionalization? An observation of current PTO activities. *Telecommunications Policy,* 17(9), 699-706.

Kuznets, S. (1953). *Economic change: Selected essays in business cycles, national income and economic growth.* New York: Norton.

Law, C. E. (1995). *Telecommunications in Eastern Europe and the CIS, markets and prospects to 2000.* London: Financial Times Management Reports.

Lazarsfeld, P. (1944). *The people's choice.* New York: Duell, Sloan and Pierce.

Leive, D. M. (1970). *International telecommunications and international law: The regulation of the radio spectrum.* Dobbs Ferry, NY: Oceana.

Lent, J. (1977). *Third world mass media and their search for modernity.* Lewisburg, PA: Bucknell University Press.

Lippmann, W. (1922). *Public opinion.* New York: The Free Press.

Mason, C. (June 20, 1994). The wireless local loop: A niche market in the United States? *Telephony.*

McDonough, P. (1981) *Power and ideology in Brazil.* Princeton, NJ: Princeton University Press.

McKnight, L., Neuman, W. R., Ferro, C., & Botelho, A. (June, 1994). *A white paper on the future of Brazil's telecommunications infrastructure.* Unpublished manuscript, Center for Science, Technology and Policy, MIT.

McLuhan, M. (1962). *The gutenberg galaxy: The making of typographical man.* Toronto: University of Toronto Press.

Milliken, M. & Blackmer, D. (1961). *The emerging nations: Their growth and United States policy.* Boston, MA: Little, Brown and Company.

Minoli, D. (1991). *Telecommunications technology handbook.* Norwood, MA: Artech House.

Mokyr, J. (1990). *The lever of riches.* Oxford: Oxford University Press.

Moss, M. L. (ed). (1980) *Telecommunications and productivity.* Reading, MA: Addison-Wesley.

Mueller, M. (July, 1989). The switchboard problem: Scale signaling and organization in manual telephone switching. 1878-1898. *Technology and Culture.*

Mueller, M. (July, 1993). Universal service in telephone history; A reconstruction. *Telecommunications Policy.* 17 (6), 367-390.

Mumford, L. (1961). *The city in history.* New York: Harcourt Brace.

Mumford, L. (1934). *Technics and civilization.* Rathway, NJ: Harcourt Brace.

Neuman, W. R. (1986). *The paradox of mass politics.* Cambridge, MA: Harvard University Press.

Neuman, W. R. (1991). *The future of the mass audience.* Cambridge: Cambridge University Press.

Neuman, W. R., Miller, F., O'Donnell, S. & Regli, B. (June, 1995). *Towards an open telecommunications environment: A white paper on the future of Russia's telecommunications networks.* Unpublished manuscript. Edward R. Murrow Center, Tufts University.

Newberg, P. R. (1989). *New directions in telecommunications policy.* Durham, SC: Duke University Press.

Newman, K. (1986). *The selling of British Telecom.* London: Holt, Rinehart and Winston.

Nisbet, R. (1966). *The sociological tradition.* New York: Basic Books.

Nisbet, R. (1980). *The history of the idea of progress.* New York: Basic Books.

Noam, E. M. (June, 1994). Beyond liberalization II: The impending doom of common carriage. *Telecommunications Policy.*

Noam, E. M. (1992). *Telecommunications in europe.* New York: Oxford University Press, 1992.

Noam, E. M. (Winter, 1987). Public telecommunications networks: A concept in transition. *Journal of Communication.*

Noam, E. M. (ed). (1983). *Telecommunications regulation today and tomorrow.* New York: Harcourt Brace.

Nordestreng, K., and Shiller, H. I. (eds). (1979). *National soverignty and international communication.* Norwood, NJ: Ablex.

NTIA Telecom 2000: Charting the course for a new century. (October, 1988). National Telecommunications and Information Administration, Department of Commerce, Government of the United States. Washington, DC: U.S. Government Printing Office.

O'Donnell, G. A., Schmitter, P. C. & Whitehead, L. (eds). (1986). *Transitions from authoritarian rule.* Baltimore, MD: Johns Hopkins University Press.

Office of Technology Assessment. (1995). *Wireless technologies and the national information infrastructure.* Washington, DC: United States Government Press.

Office of Telecommunications, The (Oftel). (1994). *OFTEL Annual Report to the Secretary of the Department of Trade and Industry (DTI).* London: Oftel.

Office of Telecommunications, The (Oftel). (1995). *OFTEL Annual Report to the Secretary of the Department of Trade and Industry (DTI).* London: Oftel.

Office of Telecommunications, The (Oftel). (1996). *OFTEL Annual Report to the Secretary of the Department of Trade and Industry (DTI).* London: Oftel.

Ong, W. (1982). *Orality and literacy, the technologizing of the word.* New York: Routledge.

O'Reilly, V. M., Hirsh, M., Defliese, P. L. & Jaenicke, H. R. (1990). *Montgomery's auditing eleventh edition.* New York: Wiley.

Organization for Economic Cooperation and Devlopment. (1992). *Regulatory reform, privatization and competition policy.* Paris: OECD Reports.

Organization for Economic Cooperation and Devlopment. (1996). *Mobile cellular communication: Pricing strategies and competition.* Paris: OECD Reports.

Petrazzini, B. A. (1995). *The political economy of telecommunications reform in developing countries.* Westport, CT: Praeger.

Pitt, D. (1980). *The telecommunications function of the British post office: A case study of bureaucratic adaptation.* Westmead, UK: Saxon House.

Pool, I. (ed). (1977). *The social impact of the telephone.* Cambridge, MA: MIT Press.

Pool, I. (1983). *Technologies of freedom.* Cambridge, MA: Harvard University Press.

Pool, I. (1990). *Technology without boundaries.* Cambridge, MA: Harvard University Press.

Porat, M. U. (1977). *The information economy.* Washington DC: US Government Printing Office.

Putnam, R. (1993). *Making democracy work: Civic traditions in modern Italy* Princeton, NJ: Princeton University Press.

Pye, R., Heath, M., Spring, G. & Yeomans, J. (February, 1991). Competition and choice in telecommunications: The duopoly review consultative document. *Telecommunications Policy.* 15 (2), 133-136.

Reed, D. P. (November, 1992). *Putting it All Together: The Cost Structure of Personal Communications Services.* Washington, DC: Federal Communications Commission, OPP Working Paper No. 28.

Renfrew, C. (1990). *Archaeology and language.* New York: Cambidge University Press.

Rogers, E. (1962). *The diffusion of innovation.* New York: The Free Press.

Rogers, E. (1994). *A history of communication study: A biographical approach.* New York: Maxwell Macmillan International.

Rosenberg, N., Landau, R., & Mowery, D.C., eds. (1993). *Technology and the wealth of nations.* Stanford, CA: Stanford University Press.

Ruggie, J. G. (1983). *The antinomies of interdependence: National welfare and the international division of labor.* New York: Columbia University Press.

Ruiz, L. K. (October, 1994). Pricing Strategies and Regulation Effects in the U.S. Cellular Telecommunications Duopolies. *Presentation to the Twenty-First Annual Telecommunications Policy Research Conference.* Solomons Island, MD.

Sanchez, M., & Corona, R. (eds). (1993). *Privatization in Latin America.* Washington, DC: The Johns Hopkins University Press.

Sartori, G. (1987). *The theory of democracy revisited.* New Jersey: Chatam House Publishers.

Saunders, R. J., Warford, J. J., & Wellenius, B. (1983). *Telecommunications and economic development.* Baltimore, MD: Johns Hopkins University Press.

Sawhney, H. (September/October, 1993). Circumventing the center; The realities of creating a telecommunications infrastructure in the USA. *Telecommunications Policy.* 15 (7), 456-475.

Schmitz, H. & Cassiolato, J. (eds). (1992). *Hi-Tech for industrial development.* New York: Routledge.

Schneider, R. M. (1991). *Order and progress: A political history of Brazil.* Boulder, CO: Westview Press.

Schumacher, E. F. (1973). *Small is beautiful.* London: Blond and Briggs.

Schumpeter, J. (1934). *The theory of economic development; an inquiry into profits, capital, credit, interest, and the business cycle.* Cambridge, MA, Harvard University Press.

Schwartz, R. E. (1996). *Wireless communications in developing countries.* Norwood, MA: Artech House.

Sen, A. (1970). *Collective choice and social welfare.* San Francisco, CA: Holden Day.

Seton-Watson, H. (1976). *Nations and States.* Bolder, CO: Westview Press.

Slack, J. D. (1984). *Communication technologies & society: Conceptions of causality and the politics of technological intervention.* Norwood, NJ: Ablex.

Skocpol, T. (1979) *The state and social revolutions.* New York: Cambridge University Press.

Smith, P. L., & Staple, G. (1994). *Telecommunications sector reform in Asia: Toward a new pragmatism.* Washington, DC: World Bank Discussion Paper No. 232.

Staple, G., & Mullins, M. (1991). *Global telecommunications traffic flows and market structures: A quantative review.* London: International Institute of Communications.

Stigler, G. (1968). *The organization of industry.* Chicago, IL: The University of Chicago Press.

Sussman, G., & Lent, G.A. (eds). (1991). *Transnational communications: Wiring the third world.* Newbury Park, CA: Sage.

Telecommunications Reports. (1992). *PCS: The quest for 2 GHz spectrum.* Washington, DC: Telecommunications Reports Publications.

Teske, P. E. (1992). *After divestiture: The political economy of state regulation.* Albany, NY: The State University of New York.

Tewes, D. (June, 1996). Dynamics of competition in the German cellular market — Unique event or model-like evolutionary path? *Presentation to the Eleventh Biennial Conference, International Telecommunications Society.* Seville, Spain.

Thompson, W. S. (1993). *The rise and fall of third world states.* Unpublished manuscript, The Fletcher School of Law and Diplomacy, Tufts University.

Todaro, M. (1981). *Economic development in the third world.* New York: Longman.

Toulan, O. (August, 1994). *Sources of local variation and their implications for strategy.* Unpublished manuscript, Alfred P. Sloan School of Management, Massachusetts Institute of Technology.

Tyler, M., & Bednarczyk, S. (August, 1994). Regulatory institutions and processes in telecommunications: An international study of alternatives. *Telecommunications Policy.*

U.S. Department of Commerce. (1991). *Telecommunications 2000.* Washington, DC: U.S. Government Printing Office.

Wagner, R. (1981). *The invention of culture.* Chicago, IL: University of Chicago Press.

Wallenstein, G. (1990). *Setting global telecommunications standards.* Norwood, MA: Artech House.

Weber, M. (1962). *Basic concepts in sociology.* London: Peter Owen Limited.

Weber, M. (1968). *The theory of social and economic organization.* New York: The Free Press.

Wellenius, B., Stern, P. A., Nulty, T. E., & Stern, R. D. (eds). (1989). *Restructuring and managing the telecommunications sector.* Washington, DC: The World Bank.

Williams, F. (1982). *The communications revolution.* London: Sage.

Williams, F. (1991). *The New Telecommunications: Infrastructure for the Information Age.* New York: Free Press.

White, L. (1962). *Medieval technology and social change.* Oxford: Oxford University Press.

White, L. (1978). *Medieval religion and technology.* Berkeley, CA: The University of California Press.

Whorf, B. L. (1956). *Language, thought and reality.* Cambridge, MA: MIT Press.

Winner, L. (1986). *The whale and the reactor.* Chicago, IL: The University of Chicago Press.

World Bank, The. (1994). *World Development Report.* Washington, DC: The World Bank.

World Bank, The. (1995). *World Development Report.* Washington, DC: The World Bank.

World Bank, The. (1996). *World Development Report.* Washington, DC: The World Bank.

Wright, D. (1993). *Broadband: Business services, technologies and strategic impact.* Norwood, MA: Artech House.

Index

A

Advanced Mobile Phone System (AMPS), 77, 125
    in Brazil 246
    in Russia 196
Andrew Corporation, 202
Argentina, 54
AT&T, 115
    and Antitrust, 111
    as US long-distance carrier, 115
    as US service monopoly, 110
    cellular, 86
Auctions, 128-138

B

Brazil, 223-258
    first telecommunications networks, 226
    impact of military on telecommunications, 228
    Brazilsat, 242
BT
    activities in Russia, 202
    market position, 157-160
    rationale for privatization, 150

C

D

E

F

Federal Communications Commission, 43

G

General Agreement on Trade and Tariffs (GATT), 195
Green, Harold, 112
Group Standard Mobile/Group Special Mobile (GSM) 78, 101

H

Hazlett, Thomas, 81
Hills, Jill, 149
Huber, Peter, 66
Hundt, Reed, 122
Hutchinson Telecom, 174

I

IDB, 200
INTELSAT, 241
International Finance Corporation (IFC), 208
International Standards Organization (ISO), 82
International Telecommunications Union (ITU), 48
Ionica, 176
ITT, 226

K

Kingsbury Commitment, The, 43

M

McCaw Cellular, 86, 124
    purchase by AT&T 116-131
McCaw, Craig, 89
MCI, 65, 115, 116, 122
Mercury Communications, 154, 158, 162

R

S

T

U

V

W

Printed in the United States
by Baker & Taylor Publisher Services